Statistical Learning Using Neural Networks

Statistical Learning Using Neural Networks

A Guide for Statisticians and Data Scientists

By

Basilio de Bragança Pereira
Federal University of Rio de Janeiro (UFRJ)
and
National Laboratory for Scientific Computing (LNCC)

Calyampudi Radhakrishna Rao
Pennsylvania State University
and
University at Buffalo

Fábio Borges de Oliveira
National Laboratory for Scientific Computing (LNCC)

CRC Press is an imprint of the
Taylor & Francis Group, an **informa** business

CRC Press
Taylor & Francis Group
6000 Broken Sound Parkway NW, Suite 300
Boca Raton, FL 33487-2742

First issued in paperback 2022

ISBN-13: 978-1-138-36450-9 (hbk)
ISBN-13: 978-1-032-33593-3 (pbk)
ISBN-13: 978-0-429-43129-6 (ebk)
DOI: 10.1201/9780429431296

Publisher's Note
The publisher has gone to great lengths to ensure the quality of this reprint but points out that some imperfections in the original copies may be apparent.

Library of Congress Cataloging-in-Publication Data

Names: Pereira, Basilio de Bragança, author. | Rao, C. Radhakrishna (Calyampudi Radhakrishna), 1920-author. | Borges de Oliveira, Fábio, author.
Title: Statistical learning using neural networks : a guide for statisticians and data scientists / by Basilio de Bragança Pereira, Calyampudi Radhakrishna Rao, Fábio Borges de Oliveira.
Description: First edition. | Boca Raton : CRC Press, 2020. | Includes bibliographical references and index.
Identifiers: LCCN 2019056045 (print) | LCCN 2019056046 (ebook) | ISBN 9781138364509 (hardback) | ISBN 9780429431296 (ebook)
Subjects: LCSH: Statistics--Data processing. | Multivariate analysis--Data processing. | Neural networks (Computer science) | Computational learning theory. | Statistics--Methodology. | Python (Computer program language)
Classification: LCC QA276.4 .P466 2020 (print) | LCC QA276.4 (ebook) | DDC 519.50285/632--dc23
LC record available at https://lccn.loc.gov/2019056045
LC ebook record available at https://lccn.loc.gov/2019056046

Visit the Taylor & Francis Web site at
http://www.taylorandfrancis.com

and the CRC Press Web site at
http://www.crcpress.com

To our families.

Contents

Preface

This book is intended to integrate two data science modeling techniques: algorithmic modeling and data modeling. This book approaches the first technique with neural networks and the second with statistical methods.

The first author's (BBP) interest in neural networks began in view of the parallel relations of neural networks and statistical methods for data analysis. Such interest motivated him to write short lecture notes on the subject for a minicourse at the SINAPE (Brazilian National Symposium in Probability and Statistics) in 1998.

The second author (CRR), in his visit to Brazil in 1999, took notice of the notes and invited BBP to visit the Statistics Department of PSU (Penn State University) for short periods in 1999 and 2000, and finally for a year in 2003, when they wrote a first version of this book as a Technical Report of the Center of Multivariate Analysis at PSU.

The interest in the technical report caught the attention of Chapman & Hall/CRC, which offered to publish the manuscript provided the authors would add the algorithmic and computational approaches, including examples. In 2018, BBP became a courtesy researcher at LNCC (Brazilian National Laboratory for Scientific Computing) and started to collaborate with the third author (FBO), who is a professor at LNCC and has research interests in computing and neural networks, especially in the areas of security and privacy.

Acknowledgments

The authors thank CAPES (Agency of the Brazilian Ministry of Education) for a year of support grant to BBP during his visit at Penn State University and to UFRJ for a leave of absence and FAPERJ (Carlos Chagas Filho Foundation for Research Support of the State of Rio de Janeiro) for a visiting grant to work on this book at the Brazilian National Laboratory for Scientific Computing (LNCC).

The authors thank Claudio Téllez and Matheus Aranha for reviewing the manuscript. The authors also thank Renata and Lawrence Hamtil for proofreading.

The authors are grateful to David Grubbs from Taylor and Francis for being a supportive and patient editor.

1

Introduction

In this chapter, we give a brief overview of statistical learning and its relation to classical statistics using as reference the two statistical cultures of Breiman: data modeling (inference) and algorithm modeling (predictive).

This book describes data analysis techniques using algorithm modeling of artificial neural networks. It also alerts statisticians to the parallel works of neural network researchers in the area of data science.

In this area of application (data science), it is important to show that most neural network models are similar or identical to popular statistical techniques such as generalized linear models, classification, cluster, projection pursuit regression, generalized additive models, principal components, factor analysis, etc. In contrast, the implementation of neural network algorithms is inefficient because:

1. They are based on biological or engineering criteria such as how easy it is to fit a net on a chip rather than on well-established statistical or optimization criteria.
2. Deep neural networks are designed to be trained on massively parallel computers. On a personal computer, standard statistical optimization criteria usually will be more efficient than neural network implementations.

In a computational process, we have the following hierarchical framework. The top of the hierarchy is the computational level. This attempts to answer the question: what is being computed and why? The next level is the algorithm, which describes how the computation is being carried out, and finally there is the implementation level, which gives the details and steps of the facilities of the algorithm.

Research in neural networks involves different groups of scientists in neurosciences, psychology, engineering, computer science, and mathematics. All these groups pose different questions: neuroscientists and psychologists want to know how the animal brain works, engineers and computer scientists want to build intelligent machines and mathematicians want to understand the fundamentals properties of networks as complex systems.

Mathematical research in neural networks is not especially concerned with statistics but mostly with nonlinear dynamics, geometry, probability and other areas of mathematics. In contrast, neural network research can stimulate statistics not only by pointing to new applications but also providing it with new type of nonlinear modeling. We believe that neural networks will become standard techniques in applied statistics, not just because biology is their source of inspiration, but also because statisticians face a range of problems whose solutions can contribute to neural network research.

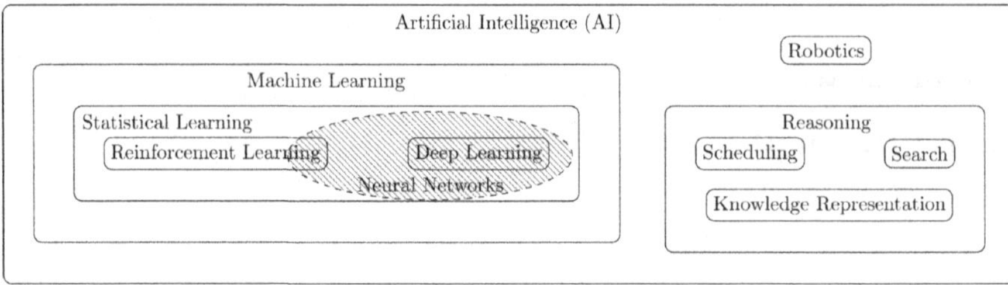

Figure 1.1
Road map for artificial intelligence.

Therefore, in this book, we will concentrate on the top of the hierarchy of the computational process, that is, what is being done and why, and how the neural network models and statistical models are related to tackle the data scientist's real problem. For further details see Gurney [82], Sarle [187], and Amari [7].

Figure 1.1 depicts the relationships of several areas of artificial intelligence (AI). Some topics in machine learning can be verified mathematically, but not all. Some important points in the mathematical foundation are undecidable [17]. The hatched area in the figure indicates the scope of this book.

The book is organized as follows: some basics on artificial neural networks (ANN) are presented in Chapters 2 and 3. Multivariate statistical topics are presented in Chapter 4. Chapter 5 deals with parametric and nonparametric regression models and Chapter 6 deals with inference results, survival analysis, control charts, and time series.

2

Fundamental Concepts on Neural Networks

This chapter deals with the fundamentals of neural networks. Symbolist and connectionist, activation functions, network architectures, Mcculloch-pitt neuron, Rosenblatt perceptron, Widrow's Adaline and Madaline and network training are described. Function approximation, statistical methods and Kolmogorov theorem in approximation are related. Model choice, terminology: neural networks versus statistical terms, and original historical types of neural networks are presented.

2.1 Artificial Intelligence: Symbolist and Connectionist

We can think of artificial intelligence as intelligent behavior embodied in human-made machines. The concept of what is intelligence lies outside the scope of this guide. It comprises the following components: learning, reasoning, problem-solving, perception, and language understanding.

From the early days of computing, there have existed two different approaches to the problem of developing machines that might embody such behavior. One of these tries to capture knowledge as a set of irreducible semantic objects or symbols and to manipulate these according to a set of formal rules. The rules taken together form a recipe or algorithm for processing the symbols. This approach is the symbolic paradigm, which can be described as consisting of three phrases:

- Choice of an intelligent activity to study.
- Development of a logic-symbolic structure able to imitate (such as "if condition 1 and condition 2... then result").
- Compare the efficiency of this structure with the real intelligent activity.

Note that the symbolic artificial intelligence is more concerned in imitating intelligence and not in explaining it. Concurrent with this, there has been another line of research, which has used machines whose architecture is loosely based on the human brain. These artificial neural networks are supposed to learn from examples and their "knowledge" is stored in representations that are distributed across of a set of weights.

The connectionist, or neural network approach, starts from the premise that intuitive knowledge cannot be captured in a set of formalized rules. It postulated that the physical structure of the brain is fundamental to emulate the brain function and the understanding of the mind. For the connectionists, the mental process comes as the aggregate result of the behavior of a great number of simple computational connected elements (neurons) that exchange signals of cooperation and competition.

The form by which these elements are interconnected is fundamental for the resulting mental process, and this is the fact that named this approach *connectionist.*

Some features of this approach are as follows:

- Information processing is not centralized and sequential but parallel and spatially distributed.
- Information is stored in the connections and adapted as new information arrives.
- It is robust to failure of some of its computational elements.

For further details see Carvalho [31].

2.2 The Brain and Neural Networks

The research work on artificial neural networks has been inspired by our knowledge of the biological nervous system, in particular the brain. The brain has other features that do not concern us here, such as brain lobes nor with lower levels of descriptions than nerve cell (neurons), and have irregular forms, as illustrated in Figure 2.1.

We will focus on the morphological the characteristic that allows the neuron to function as an information processing device. This characteristic lies in the set of fibers that emanate from the cell body. One of these fibers - the axon - is responsible for transmitting information to other neurons. All others are dendrites, which carry information transmitted from other neurons. Dendrites are surrounded by the synaptic boutons of other neurons, as in Figure 2.1(c). Neurons are highly interconnected. Physical details are given in Table 2.1 where we can see that the brain is a very complicated system, and a true model of the brain would be very complicated. Scientists and medical professionals require detailed models. However, statisticians and engineers have found that simpler models can be built easily and manipulated effectively. As statisticians, we will take the view that brain models are used as inspiration to building artificial neural networks as the wings of a bird were the inspiration for the wings of an airplane. Table 2.1 also presents two comparisons using the processor IBM zEC12 as reference.

For more details, see Wilde [223], Gurney [82], and Carvalho [31].

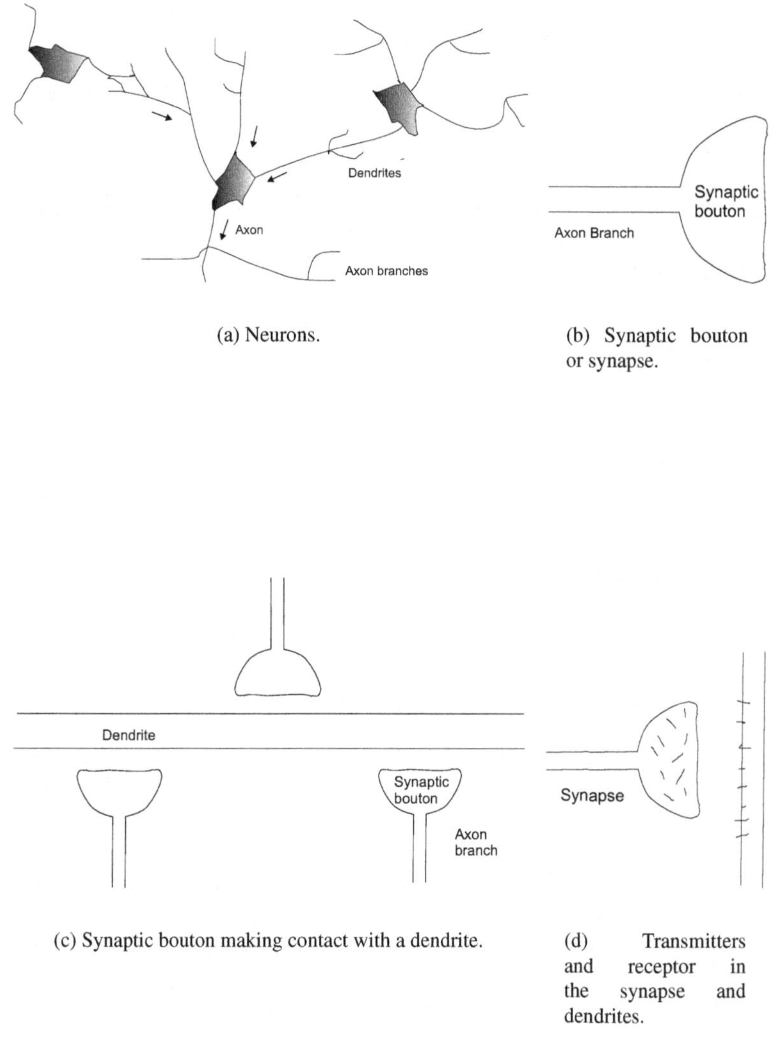

(a) Neurons.

(b) Synaptic bouton or synapse.

(c) Synaptic bouton making contact with a dendrite.

(d) Transmitters and receptor in the synapse and dendrites.

Figure 2.1
Neural network elements.

2.3 Artificial Neural Networks and Diagrams

An artificial neuron can be identified by three basic elements:

1. A set of synapses, each characterized by a weight, that, when positive, means that the synapse is excited, and that, when negative, is inhibitory. A signal x_j

Table 2.1
Some physical characteristics of the brain.

Number of neurons	86 billion
Total number of synapses (number of transistors of main processor)	602 000 billion (more than 2.7 billion)
Number of synapses/neurons	7 000
Operation frequency (processor frequency)	Average up to 100 hertz (more than 5 billion, 5 gigahertz)
Human brain volume	1 200 cm^3
Dendrite length of a neuron	2 μm
Brain weight	1.4 kg
Neuron weight	1.2×10^{-9} g
Synapse	Excitatory and inhibitory

in the input of synapse j connected to the neuron k is multiplied by a synaptic weight w_{jk}.

2. Summation - to sum the input signs, weighted by respective synapses (weights).
3. Activation function - restrict the amplitude of the neuron output.

The Figure 2.2 presents a model of an artificial neuron displayed as a network diagram.

- Neurons are represented by circles and boxes, while the connections between neurons are shown as arrows.
- Circles represent observed variables, with the names shown inside the circles.
- Boxes represent values computed as a function of one or more arguments. The symbol inside the box indicates the type of function.
- Arrows indicate that the source of the arrow is an argument of the function computed at the destination of the arrow. Each arrow usually has a weight or parameter to be estimated.
- A thick line indicates that the values at each end are to be fitted by the same criterion such as least squares, maximum likelihood etc.

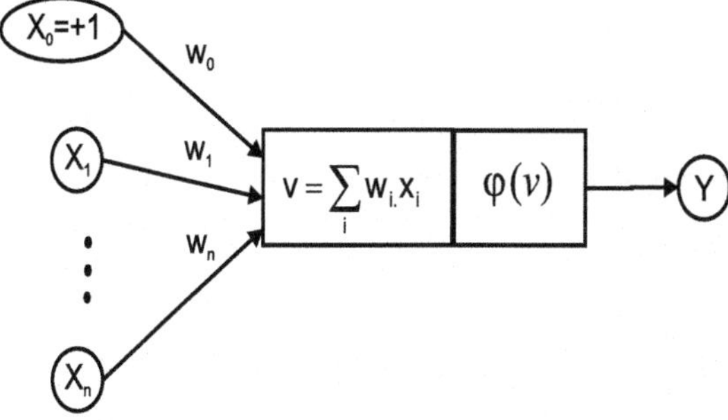

Figure 2.2
Model of an artificial neuron.

- An artificial neural network is formed by several artificial neurons, based on the arrangements of biological neurons.
- Each artificial neuron is constituted by one or more inputs and one output. These inputs might be outputs of other neurons and the output might be the input to other neurons. The inputs are multiplied by weights and summed with one constant; this total then goes through the activation function. This function has the objective of activating or inhibiting the next neuron.
- Mathematically, we describe the kth neuron with the following form:

$$u_k = \sum_{j=1}^{p} w_{kj} x_j \text{ and } y_k = \varphi(u_k - w_{k0}), \tag{2.1}$$

where $x_0, x_1, \ldots, x_p$ are the inputs; $w_{k1}, \ldots, w_{kp}$ are the synaptic weights; u_k is the output of the linear combination; w_{k0} is the bias; $\varphi(\cdot)$ is the activation function and y_k is the neuron output.

The first layer of an artificial neural network, called *input layer*, is constituted by the inputs x_p and the last layer is the *output layer*. The intermediary layers are called *hidden layers*.

The number of layers and the quantity of neurons in each one of them are determined by the nature of the problem.

A vector of values presented once to all the output units of an artificial neural network is called *case, example, pattern, sample*, etc. For more details, see Sarle [187] and Haykin [86].

2.4 Activation Functions

An activation function for a neuron can be of several types. The most common types are presented in Figure 2.3.

The derivative of the identity function is 1. The derivative of the step function is not defined, which is why it is not used.

Sigmoid functions such as in Figure 2.3d and 2.3e are easy to compute. The derivative of the logistic is $\varphi_1(x)(1 - \varphi_1(x))$.

For the hyperbolic tangent, the derivative is

$$1 - \varphi_3^2(x). \tag{2.2}$$

See also Haykin [86], Iyengar et al. [99], and Orr [156].

2.5 Network Architectures

Architecture refers to the manner in which neurons in a neural network are organized. There are three different classes of network architectures (see Haykin [86]).

1. **Single-Layer Feedforward Networks**

   In a layered neural network (Figure 2.4), the neurons are organized in layers. One input layer of *source nodes* projects onto an *output layer* of neurons, but not vice-versa. This network is strictly of a feedforward type.

2. **Multilayer Feedforward Networks**

   These are networks with one or more *hidden layers* (Figure 2.5). Neurons in each of the layers have as their inputs the output of the neurons of the preceding layer only.

   This type of architecture covers most of the statistical applications. The neural network is said to be fully connected if every neuron in each layer is connected to each node in the adjacent forward layer; otherwise, it is partially connected.

3. **Recurrent Network**

   In a recurrent network, at least one neuron connects with another neuron of the preceding layer and creates a feedback loop (Figure 2.6).

   This type of architecture is mostly used in optimization problems.

$$y = \varphi(x) = x$$

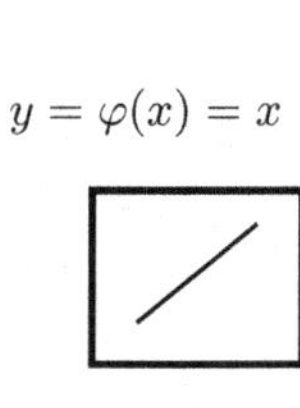

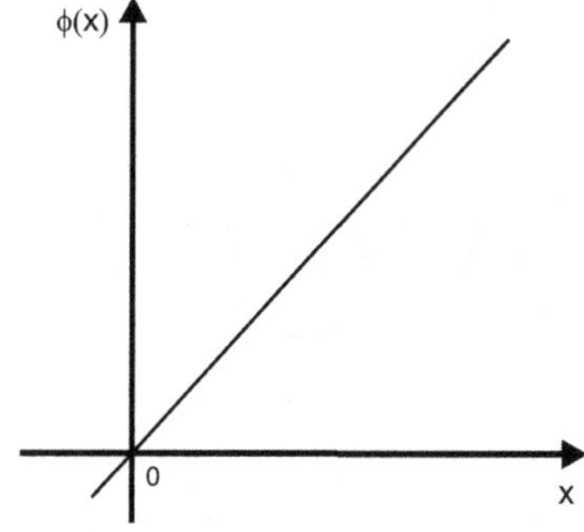

a) Identity function (ramp).

$$y = \varphi(x) = \begin{cases} 1 & x \geq b \\ -1 & x < b \end{cases}$$

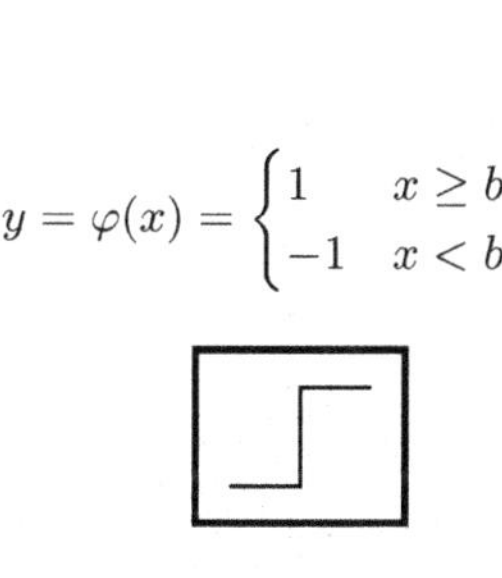

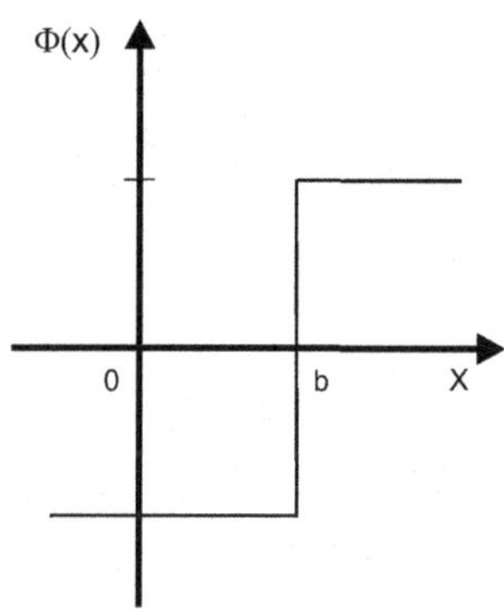

b) Step or threshold function.

$$y = \varphi(x) = \begin{cases} 1 & x \geq 1/2 \\ z & -1/2 \leq x < \frac{1}{2} \\ 0 & x \leq -\frac{1}{2} \end{cases}$$

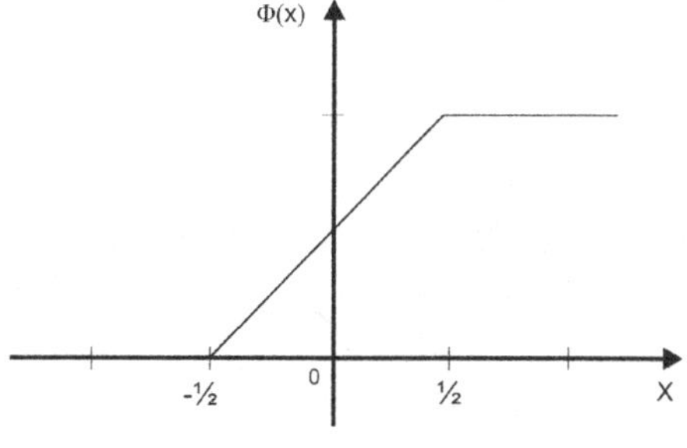

c) Piecewise-linear function.

Figure 2.3
Activation functions: graphs and ANN representations (continued).

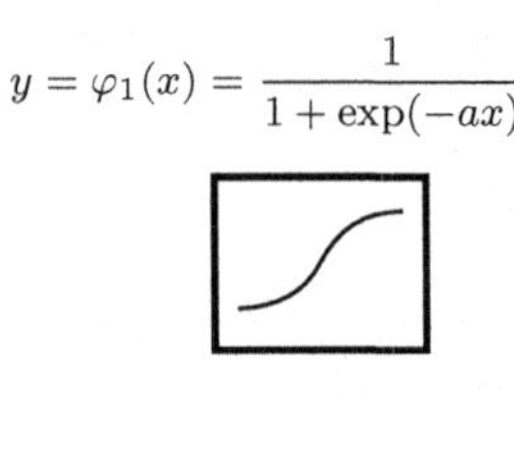

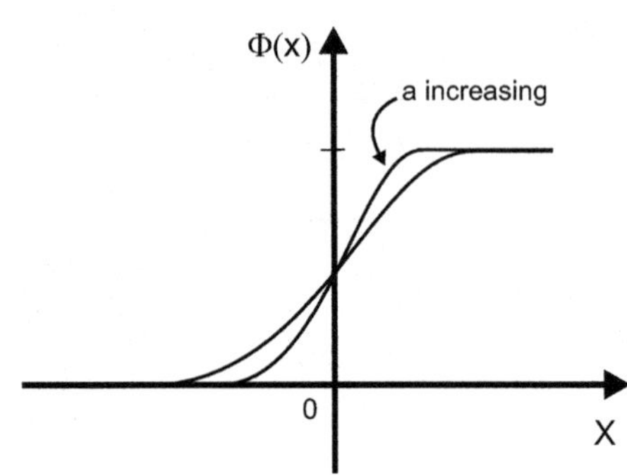

d) Logistic (sigmoid).

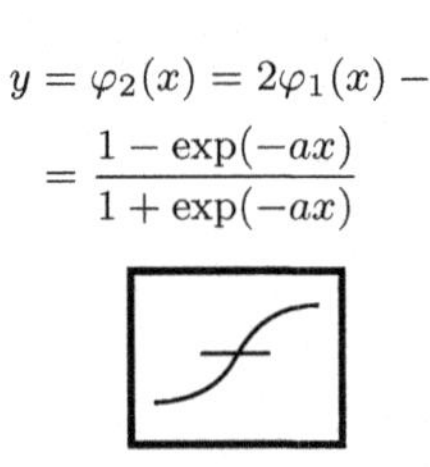

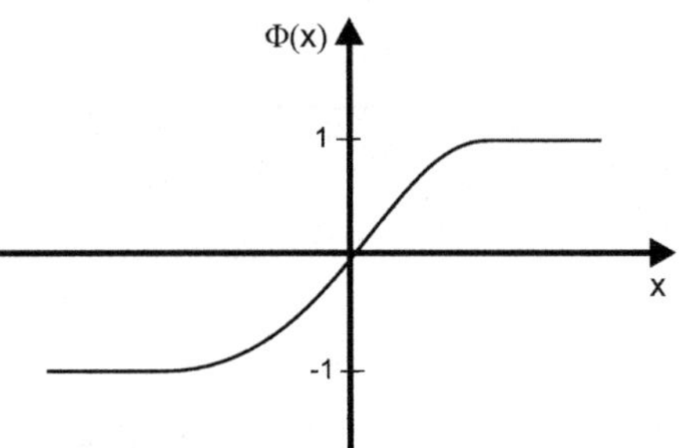

e) Symmetrical sigmoid.

$$y = \varphi_3(x) = \tanh\left(\frac{x}{2}\right) = \varphi_2(x)$$

f) Hyperbolic tangent function.

Figure 2.3
(Continued)

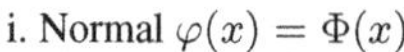

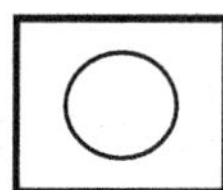

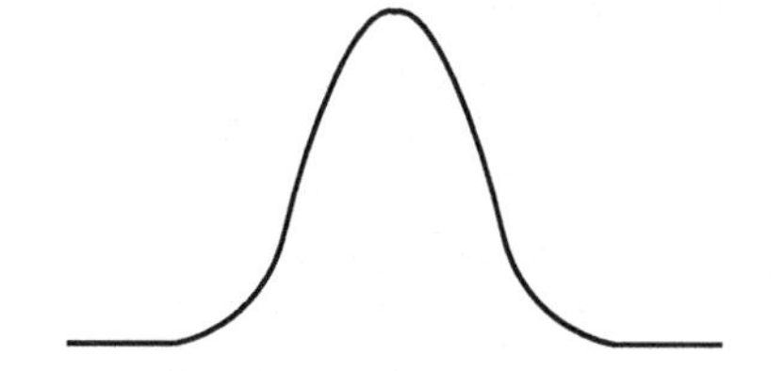

ii. Wavelet $\varphi(x) = \Psi(x)$

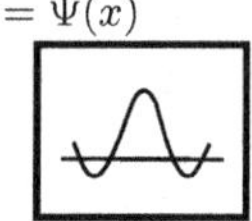

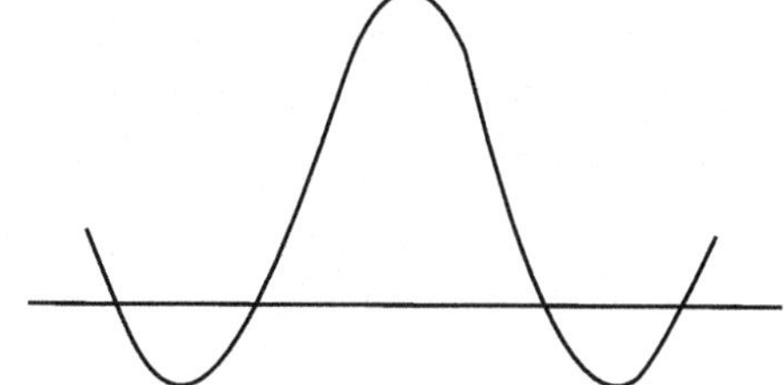

g) Radial basis (some examples).

$$\varphi(x) = \sum \sum 2^{j/2} \Psi \left(2^j x - k\right)$$

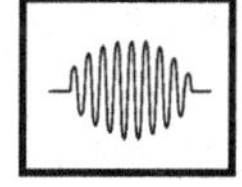

h) Wavelet transform.

$$y = \varphi(x) = \max(x, 0) = \begin{cases} x & x \geq 0 \\ 0 & x < 0 \end{cases}$$

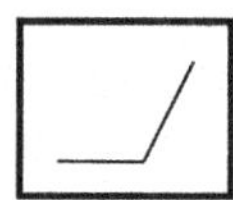

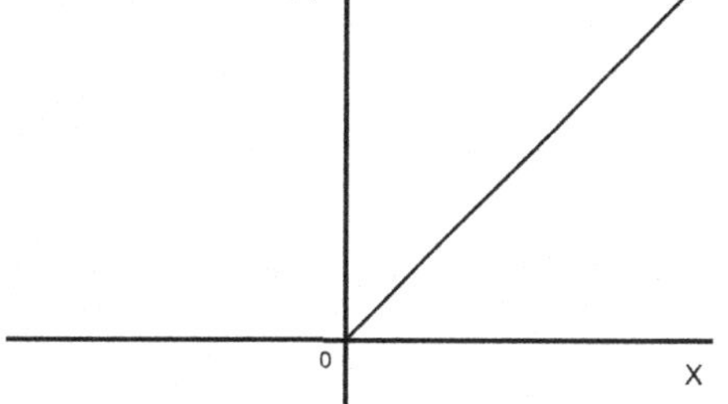

i) Rectified linear unit (ReLU) function.

Figure 2.3

2.6 Network Training

The training of the network consists of adjusting the synapses or weight with the objective of optimizing the performance of the network. From a statistical point of view, the training corresponds to estimating the parameters of the model given a set of data and an estimation criterion.

The training can be as follows:

- Supervised learning is learning with a teacher. In this case, each input vector and its output are known, so we can modify the network synaptic weights in an orderly way to achieve the desired objective.
- Unsupervised learning is learning without a teacher. In this case, we do not have the output, which corresponds to each input vector.
  - Self-organized training corresponds, for example, to the statistical methods of cluster analysis and principal components.
  - Reinforced learning training corresponds to, for example, the credit-assign problem. It is the problem of assigning a credit or a blame for overall outcomes to each of the internal decisions made by a learning machine.

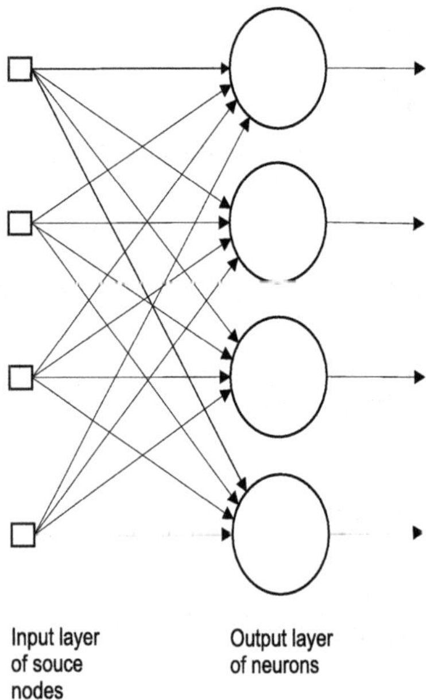

Figure 2.4
Single-layer network.

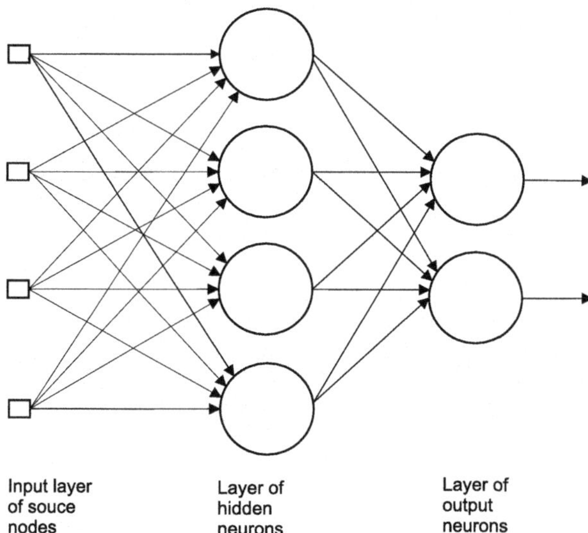

Figure 2.5
Multilayer network.

2.7 Kolmogorov Theorem

Most conventional multivariate data analysis procedures can be modeled as a mapping

$$f : A \to B$$

where A and B are finite sets. Consider the following examples in Murtagh [148]:

(i) B is a set of discrete categories, and A a set of m-dimensional vectors characterizing n observations. The mapping f is a clustering function and choosing the appropriate f is the domain of cluster analysis.

(ii) In dimensionality-reduction technique (PCA, FA, etc.), B is a space containing n points, which has a lower dimension than A. In these cases, the problem of the data analyst is to find the mapping f, without precise knowledge of B.

(iii) When precise knowledge of B is available, B has lower dimension than A, and f is to be determined by methods such as discrimination analysis and regression.

Most of the neural network algorithm can also be described as method to determine a nonlinear f, where $f : A \to B$ where A is a set of n, m-dimensional descriptive vectors and where some members of B are known a priori. Also, some

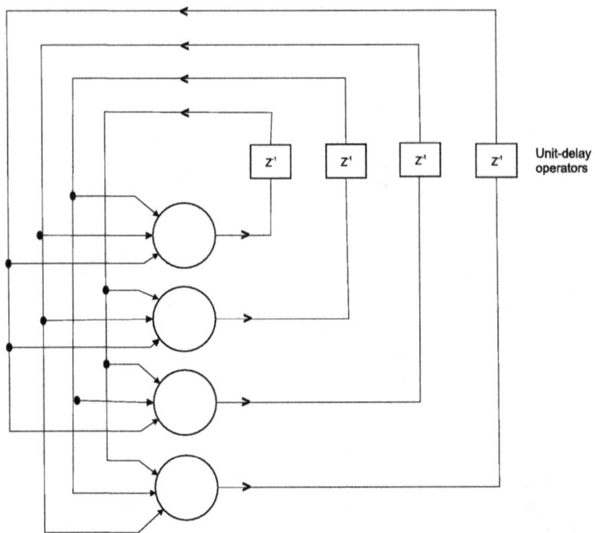

Figure 2.6
Recurrent network.

of the most common neural networks are said to perform similar tasks of statistical methods, for example, multilayer perceptron and regression analysis, Kohonen network and multidimensional scaling, adaptive resonance theory (ART), and cluster analysis.

Therefore, the determination of f is of utmost importance and made possible due to a theorem of Andrei Kolmogorov about representation and approximation of continuous functions, whose history and development we present shortly.

In 1900, Professor David Hilbert, at the International Congress of Mathematicians in Paris, formulated 23 problems from the discussion of which "advancements of science may be expected". Some of these problems have been related to applications, where insights into the existence, but not the construction, of solutions to problems have been deduced.

The thirteenth problem of Hilbert is having some impact in the area of neural networks, since its solution in 1957 by Kolmogorov. A modern variant of Kolmogorov's Theorem is given in Rojas [180, p. 265] (or Bose and Liang [24, p. 158]).

Theorem 2.1 (Kolmogorov Theorem) *"Any continuous function $f(x_1, x_2, \ldots, x_m)$ on n variables $x_1, x_2, \ldots, x_m$ on $I_n = [0,1]^m$ can be represented in the form*

$$f(x_1, \ldots, x_m) = \sum_{q=1}^{2m+1} g\left(\sum_{p=1}^{m} \lambda_p h_q(x_p)\right)$$

where g and h_q, for $q = 1, \ldots, 2m+1$ are functions of one variable and λ_p for $p = 1, \ldots, m$ are constants, and the g_q functions do not depend on $f(x_1, \ldots, x_m)$."

Kolmogorov Theorem states an exact representation for a continuous function but does not indicate how to obtain those functions. However, if we want to approximate a function, we do not demand exact reproducibility but only a bounded approximate error. This is the approach taken in neural networks.

From Kolmogorov's work, several authors have improved the representation to allow sigmoid functions, Fourier, radial basis, and wavelets functions to be used in the approximations.

Some of the adaptations of Kolmogorov to neural networks are in Mitchell [145, p. 105].

- **Boolean functions**. Every Boolean function can be represented exactly by some network with two layers of neurons, although the number of hidden neurons required grows exponentially in the worst case with the number of network inputs.
- **Continuous functions**. Every bounded continuous function can be approximated with arbitrary small error by a network with two layers. The result in this case applies to networks that use sigmoid activation at the hidden layer and (unthresholded) linear activation at the output layer. The number of hidden neurons required depends on the function to be approximated.
- **Arbitrary functions**. Any function can be approximated to arbitrary accuracy by a network with three layers of neurons. Again, the output layer uses linear activation, the two hidden layers use sigmoid activation, and the number of neurons required in each hidden layer also depends on the function to be approximated.

2.8 Model Choice

We have seen that the data scientist's interest is to find or learn a function f from A to B. In some cases (dimension reduction, cluster analysis), we do not have precise knowledge of B and the term "unsupervised" learning is used. In contrast, the term "supervised" is often applied to the situation where f is to be determined, but where some knowledge of B is available (e.g., discriminant and regression analysis).

In the case of supervised learning the algorithm to be used in the determination of f embraces the phases: learning, where the mapping f is determined using the training set, $B' \subset B$; testing of how well f performs using a test set $B'' \subset B$, B'' not identical to B'; and an application. The testing and application phases are named generalization.

In a multilayer network, Kolmogorov's results provide guidelines for the number of layers to be used. However, some problems may be easier to solve using two hidden layers. The number of units in each hidden layer and (the number) of training interactions to estimate the weights of the network have also to be specified. In this section, we will address each of these problems.

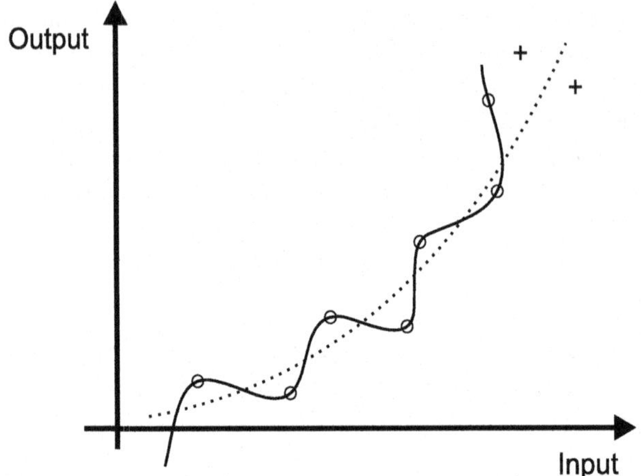

Figure 2.7
Network fits: the solid line indicates high variance, and the dotted line indicates high bias.

2.8.1 Generalization

By generalization, we mean that the input-output mapping computed by the network is correct for unused test data. It is assumed that the training will utilize a representative sample from the environment where the network will be applied.

Given a large network, it is possible that loading data into a network may cause the system to memorize the training data and insert a characteristic such as noise in the data that is not present in the underlying function to be modeled.

The training of a neural network may be viewed as a curve fitting problem and if we keep improving our fitting, we end up with a very good fit only for the training data. As a result, the neural network will not be able to generalize (interpolate or predict other sample values).

2.8.2 Bias-variance Trade-off: Early Stopping Method of Training

Consider the two functions in Figure 2.7 obtained from two trainings of a network.

The data was generated from function $h(x)$, and the data from

$$y(x) = h(x) + \epsilon. \tag{2.3}$$

The training data are shown as circles (○) and the new data by plus signs (+). The dashed curve represents a simple model that does not do a good job in fitting the new data. We say that the model has a bias. The full line represents a more complex model, which does an excellent job in fitting the data, as the error is close to zero. However, it would not do a good job for predicting the values of the test data (or new

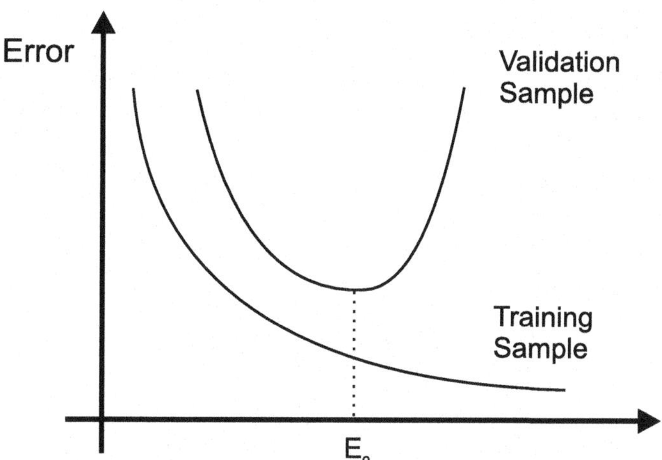

Figure 2.8
Model choice plotting training against validation.

values). We say that the model has high variance. We say in this case that the network has memorized the data or is overfitting the data.

A simple procedure to avoid overfitting is to divide the training data in two sets, namely an estimation set and a validation set. Every now and then, stop training and test the network performance in the validation set, with no weight update during this test. Network performance in the training sample will improve as long as the network is learning the underlying structure of the data (that is $h(s)$). Once the network stops learning things that are expected to be true of any data sample and learns things that are true only of the training data (that is ϵ), performance on the validation set will stop improving and will get worse.

Figure 2.8 shows the errors on the training and validation sets. Each training period is known as an epoch. To stop overfitting, we stop training at epoch E_0.

Haykin [85, p. 215] suggests taking 80% of the training set for estimation and 20% for validation.

2.8.3 Choice of Structure

A large number of layers and neurons reduces bias but at the same time increases variance. In practice it is usual to try several types of neural networks and to use cross validation on the behavior in test data, to choose the simplest network with a good fit.

We may achieve this objective in one of two ways:

- Network growing;
- Network pruning.

2.8.4 Network Growing

One possible policy for the choice of the number of layers in a multilayer network is to consider a nested sequence of networks of increasing sequence as suggested by Haykin [85, p. 215].

- p networks with a single layer of increasing size $h'_1 < h'_2 < \ldots < h'_p$.
- g networks with two hidden layers: the first layer of size h_p and the second layer is of increasing size $h''_1 < h''_2 < \ldots < h''_g$.

The cross validation approach similar to the early stopping method previously presented can be used for the choice of the numbers of layers and also of neurons.

2.8.5 Network Pruning

In designing a multilayer perceptron, we are building a nonlinear model of the phenomenon responsible for the generation of the input-output examples used to train the network. As in statistics, we need a measure of the fit between the model (network) and the observed data.

Therefore, the performance measure should include a criterion for selecting the model structure (number of parameters of the network). Various identification or selection criteria are described in the statistical literature. The names of Akaike, Hannan, Rissanen, and Schwartz associated with these methods share a common form of composition:

$$\begin{pmatrix}\text{Model-Complexity}\\ \text{criteria}\end{pmatrix} = \begin{pmatrix}\text{Performance}\\ \text{measure}\end{pmatrix} + \begin{pmatrix}\text{Model-Complexity}\\ \text{penalty}\end{pmatrix}$$

The basic difference between the various criteria lies in the model-complexity penalty term.

In the context of backpropagation learning or other supervised learning procedure the learning objective is to find a weight vector that minimizes the total risk

$$R(W) = E_s(W) + \lambda E_c(W) \tag{2.4}$$

where $E_s(W)$ is a standard performance measure that depends on both the network and the input data. In backpropagation learning, it is usually defined as mean-square error. In addition, $E_c(W)$ is a complexity penalty that depends on the model alone.

The regularization parameter λ represents the relative importance of the complexity term with respect to the performance measure. With $\lambda = 0$, the network is completely determined by the training sample. When λ is large, it says that the training sample is unreliable.

Note that this risk function, with the mean square error, is closely related to ridge regression in statistics and the work of Tikhonov on ill posed problems or regularization theory (see Orr [156] and Haykin [85, p. 219]).

A complexity penalty used in neural networks described by Haykin [85, p. 220] is

$$E\{w\} = \sum_i \frac{(w_i/w_o)^2}{1 + (w_i/w_o)^2}, \tag{2.5}$$

where w_0 is a preassigned parameter. Using this criteria, weight's w_i's with small values should be eliminated.

Another approach to eliminate connections in the network, works directly in the elimination of units and was proposed and applied by Park et al. [159]. It consists in applying PCA, principal components analysis as follows:

(i) initially train a network with an arbitrary large number P of hidden modes;

(ii) apply PCA to the covariance matrix of the outputs of the hidden layer;

(iii) choose the number of important components P^* (by one of the usual criteria: eigenvalues greater than one, proportion of variance explained etc.);

(iv) if $P^* < P$, pick P^* nodes out of P by examining the correlation between the hidden modes and the selected principal components.

Note that the resulting network will not always have the optimal number of units. The network may improve its performance with few more units.

2.9 McCulloch-Pitt Neuron

The McCulloch-Pitt neuron is a computational unit of binary threshold, also called Linear Threshold Unit (LTU). The neuron receives the weighted sum of inputs and has as output the value 1 if the sum is bigger than this threshold and it is shown in Figure 2.9.

Mathematically, we represent this model as

$$y_k = \varphi\left(\sum_{j=1}^{p} \omega_{kj} - \omega_o\right), \tag{2.6}$$

where y_k is an output of neuron k, ω_{jk} is the weight of neuron j to neuron k. x_j is the output of neuron j, ω_o is the threshold to the neuron k and φ is the activation function defined as

$$\varphi(\text{input}) = \begin{cases} 1 & \text{if input} \geq \omega_0 \\ -1 & \text{otherwise} \end{cases}. \tag{2.7}$$

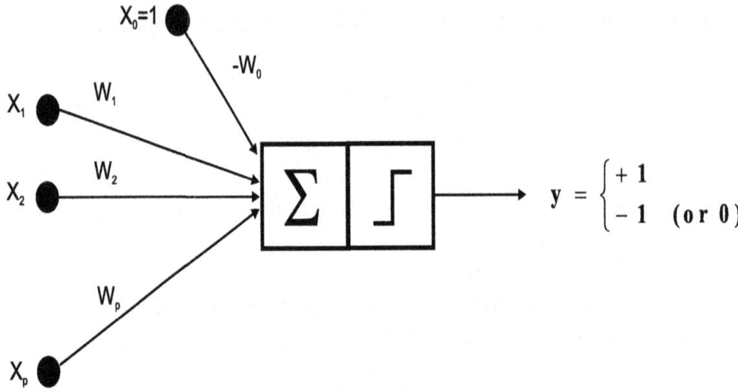

Figure 2.9
McCulloch-Pitt neuron.

2.10 Rosenblatt Perceptron

The Rosenblatt perceptron is a model with several McCulloch-Pitt neurons. Figure 2.10 shows one such model with two layers of neurons.

The one-layer perceptron is equivalent to a linear discriminant:

$$\sum \omega_j x_j - \omega_0. \tag{2.8}$$

This perceptron creates an output indicating if it belongs to class either 1 or -1 (or 0). That is

$$y = \text{Out} = \begin{cases} 1 & \sum \omega_i x_i - \omega_0 \geq 0 \\ 0 & \text{otherwise.} \end{cases} \tag{2.9}$$

The constant ω_0 is referred as threshold or bias in computational jargon. In the perceptron, the weights are constraints and the variables are the inputs. Since the objective is to estimate (or learn) the optimum weights, as in other linear discriminants, the next step is to train the perceptron.

The perceptron is trained with samples using a sequential learning procedure to determine the weight as follows: The samples are presented sequentially, for each wrong output the weights are adjusted to correct the error. If the output is corrected the weights are not adjusted. The equations for the procedure are:

$$\omega_i(t+1) = \omega_i(t) + \Delta\omega_i(t) \tag{2.10}$$

$$\omega_0(t+1) = \omega_0(t) + \Delta\omega_0(t) \tag{2.11}$$

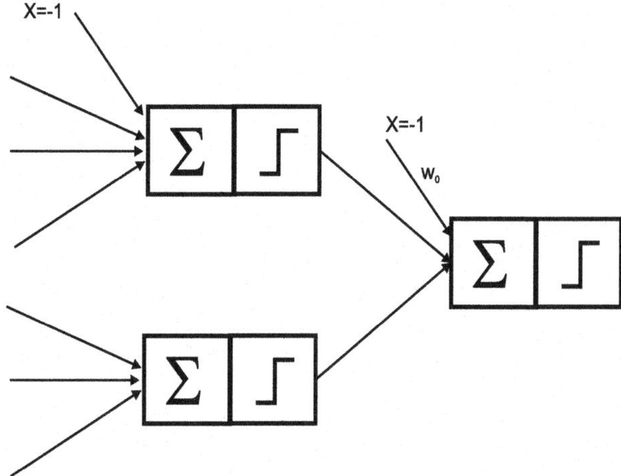

Figure 2.10
Two-layer perceptron.

and

$$\Delta\omega_i(t) = (y - \hat{y})x_i \tag{2.12}$$
$$\Delta\omega_0(t) = (y - \hat{y}). \tag{2.13}$$

Formally, the actual weight in true t is $\omega_i(t)$, the new weight is $\omega_i(t+1)$, $\Delta\omega_i(t)$ is the adjustment or correction factor, y is the correct value or true response, $\hat{y}$ is the perceptron output, a is the activation value for the output $\hat{y} = f(a)$ ($a = \sum \omega_i x_i$).

The initial weight values are generally random numbers in the open interval $(0, 1)$. The sequential presentation of the samples continues indefinitely until some stop criterion is satisfied. For example, if 100 cases are presented and if there is no error, the training will be completed. Nevertheless, if there is an error, all 100 cases are presented again to the perceptron. Each cycle of presentation is called an epoch. To make learning and convergence faster, some modifications can be done in the algorithm. For example,

(i) To normalize the data: all input variables must be within the open interval $(0, 1)$.

(ii) To introduce a *learning rate* α in the weight actualization procedure to $\alpha\omega_i$. The value of α is a number in the open interval $(0, 1)$.

The perceptron convergence theorem states that when two classes are linearly separable, the training guarantees the convergence but not its speed. Thus, if a linear discriminant exists and can separate the classes without committing an error, the training procedure will find the line or separator plan, but it is not known how long it will take to find it (Figure 2.11).

Table 2.2 illustrates the computation of results of this section for the example of Figure 2.11. Here convergence is achieved in one epoch. Here $\alpha = 0.25$.

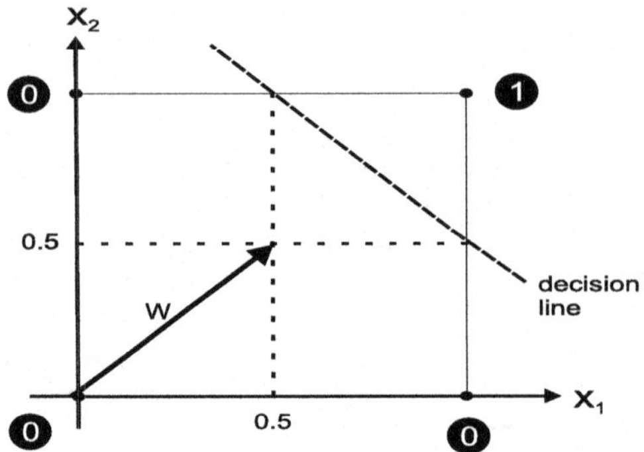

Figure 2.11
Two-dimensional example.

Table 2.2
Training with the perceptron rule on a two-input example.

ω_1	ω_2	ω_0	x_1	x_2	a	y	$\hat{y}$	$a(\hat{y}-y)$	$\Delta\omega_1$	$\Delta\omega_2$	$\Delta\omega_0$
0.0	0.4	0.3	0	0	0	0	0	0	0	0	0
0.0	0.4	0.3	0	1	0.4	1	0	-0.25	0	-0.25	0.25
0.0	0.15	0.55	1	0	0	0	0	0	0	0	0
0.0	0.15	0.55	1	1	0.15	0	1	0.25	0.25	0.25	-0.25

2.11 Widrow's Adaline and Madaline

Another neuron model was developed by Widrow and Hoff and called ADALINE (ADAptive LINear Element), which uses as a training the method of Least Mean Square (LMS) error. Its generalization consists of a structure of many ADALINEs, namely MADALINE.

The major difference between the perceptron (or LTU) and the ADALINE is in the activation function. While the perceptron uses the step or threshold function, the ADALINE uses the identity function as in Figure 2.12.

To a set of samples, the performance measure of the training is

$$D_{MLS} = \sum (\tilde{y} - y)^2. \tag{2.14}$$

The LMS training objective is to feed the weights that minimize D_{MLS}.

A comparison of training for the ADALINE and perceptron for a two dimensional neuron with $w_0 = -1$, $w_1 = 0.5$, $w_2 = 0.3$ and $x_1 = 2$, $x_2 = 1$ would result in an

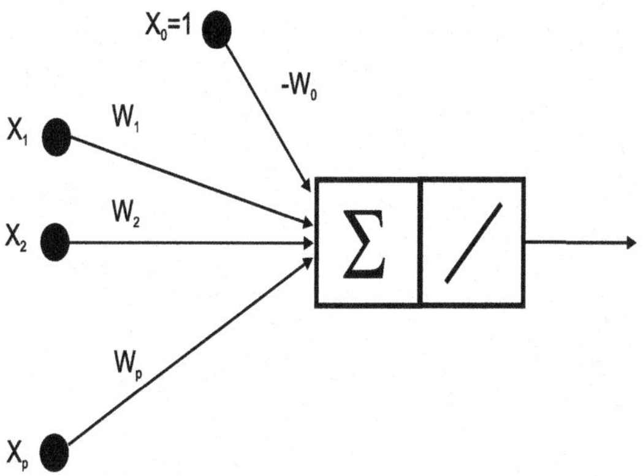

Figure 2.12
ADALINE.

Table 2.3
Comparison of updating weights.

	Perceptron	Adaline
$\tilde{y}$	1	0.3
$w_1(t+1)$	$0.5 + (0-1).2 = -1.5$	$0.5 + (0-0.3) = -0.1$
$w_2(t+1)$	$0.3 + (0-1).2 = -0.7$	$0.3 + (0-0.3).1 = 0$
$w_0(t+1)$	$-1 + (0-1) = -2$	$-1 + (0-0.3) = -1.3$

activation value of:

$$a = 2 \times 0.5 + 1 \times 0.3 - 1 = 0.3. \tag{2.15}$$

Table 2.3 gives the updating results for the sample case.

The technique used to update the weight is called gradient descent. The perceptron and the LMS search for a linear separator. The appropriate line or plane can only be obtained if the classes are linear separable. To overcome this limitation the more general neural network model of the next chapter can be used.

2.12 Terminology

Although many neural networks are similar to statistical models, a different terminology has been developed in neurocomputing. Table 2.4 helps to translate from the neural network language to statistics [188]. Some further glossary lists are given in the references for this section. Some are quite specialized, for instance, we suggest Addrians and Zanringe [5] for statistic and data mining terminology and Gurney [82] for statisticians and neuroscientists.

Table 2.4
Terminology

Neural Network Jargon	Statistical Jargon
Generalizing from noisy data and assessment of accuracy thereof	Statistical inference
The set of all cases one wants to be able to generalize to	Population
A function of the value in a population, such as the mean or a globally optimal synaptic weight	Parameter
A function of the values in a sample, such as the mean or a learned synaptic weight	Statistic
Neuron, neurocode, unit, node processing element	A simple linear or nonlinear computing element that accepts one or more inputs, computes a function thereof, and may direct the result to one or more other neurons
Neural networks	A class of flexible nonlinear regression and discriminant models, data reduction models, and nonlinear dynamical systems consisting of an often large number of neurons interconnected in often complex ways and often organized in layers
Statistical methods	Linear regression and discriminant analysis, simulated annealing, random search
Architecture	Model
Training, learning, adaptation	Estimation, model fitting, optimization
Classification	Discriminant analysis
Mapping, function approximation	Regression
Supervised learning	Regression, discriminant analysis

Continued on next page

Neural Network Jargon	Statistical Jargon
Unsupervised learning, self-organization	Principal components, cluster analysis, Data reduction
Competitive learning	Cluster analysis
Hebbian learning, Cottrell-Munzo-Zipser technique	Principal components
Training set	Sample, construction sample
Test set, validation set	Hold-out sample
Pattern, vector, example, case	Observation, case
Binary (0/1), bivalent or bipolar (-1/1)	Binary, dichotomous
Input	Independent variables, predictors, regressors, explanatory variables, carriers
Output	Predicted values
Forward propagation	Prediction
Training values	Dependent variables, responses
Target values	Observed values
Training par	Observation containing both inputs and target values
Shift register, (tapped) (time) delay (line), Input window	Lagged value
Errors	Residuals
Noise	Error term
Generalization	Interpolation, extrapolation, prediction
Error bars	Confidence intervals
Prediction	Forecasting
ADALINE (ADAptive LInear NEuron)	Linear two-group discriminant analysis (not a Fisher's but generic)
(No-hidden-layer) perceptron	Generalized linear model (GLIM)

Continued on next page

Neural Network Jargon	Statistical Jargon
Activation function, signal function, Transfer function	Inverse link function in GLIM
Softmax	Multiple logistic function
Squashing function	Bounded function with infinitive domain
Semilinear function	Differentiable nondecreasing function
Phi-machine	Linear model
Linear 1-hidden-layer perceptron	Maximum redundant analysis, principal components of instrumental variables
1-hidden-layer perceptron	Projection pursuit regression
Weights synaptic weights	(Regression coefficients, parameter estimates)
Bias	Intercept
The difference between the expected value of a statistic and the corresponding true value (parameter)	Bias
OLS (orthogonal least square)	Forward stepwise regression
Probabilistic neural network	Kernel discriminant analysis
General regression neural network	Kernel regression
Topolically distributed enconding	(Generalized) additive model
Adaptive vector quantization	Iterative algorithm of doubtful convergence for K-means cluster analysis
Adaptive resonance theory $2^n d$	Hartigan's leader algorithm
Learning vector quantization	A form of piecewise linear discriminant analysis using a preliminary cluster analysis
Counterpropagation	Regressogram based on k-means clusters
Encoding, Autoassociation	Dimensionality reduction (independent and dependent variables are the same)

Continued on next page

Neural Network Jargon	Statistical Jargon
Heteroassociation	Regression, discriminant analysis (independent and dependent variables are different)
Epoch	Iteration
Continuous training, incremental training, on-line training, Instantaneous training	Iteratively updating estimates one observation at time via difference equation, as in stochastic approximation
Batch training, off-line training	Iteratively updating estimates after each complete pass over the data as in most nonlinear regression algorithms
Shortcuts, jumpers, bypass connections, direct linear feedthrough (direct connections from input to output)	Main effects
Funcional links	Iteration terms or transformations
Second-order network	Quadratic regression, response-surface model
Higher-order networks	Polynomial regression, linear model with iteration terms
Instar, outstar	Iterative algorithms of doubtful convergence for approximating an arithmetic mean or centroid
Delta rule, ADALINE rule, Widrow-Hoff rule, LMS rule	Iterative algorithm of doubtful convergence for training a linear perceptron by least squares, similar to stochastic approximation
LMS (least mean squares)	OLS (ordinary least squares) (see LMS rule above)
Training by minimizing the median of the squared errors	LMS (least median of square)
Generalized delta rule	Iterative algorithm of doubtful convergence for training a nonlinear perceptron by least squares, similar to stochastic approximation

Continued on next page

Neural Network Jargon	Statistical Jargon
Backpropagation	Computation of derivatives for a multi-layer perceptron and various algorithms (such as generalized delta rule) based thereon
Weight decay, regularization	Shrinkage estimation, ridge regression
Jitter	Random noise added to the inputs to smooth the estimates
Growing, pruning, brain damage, self-structuring, ontogeny	Subset selection, model selection, pre-test estimation
Optimal brain surgeon	Wald test
Relative entropy, cross entropy	Kullback-Leibler divergence
Evidence framework	Empirical Bayes estimation

For further details see Gurney [82], Kasabov [109], Addrians and Zanringe [5], Fausett [69], Stegemann and Buenfeld [198], Sarle [187], Sarle [188], and Arminger and Enache [8].

2.13 Running Python in a Nutshell

Python is a very flexible programming language available for several operating systems. Indeed, some of them already have Python installed. If you need to install it, you might use a package manager in your operating system, for instance, as follows.

```
sudo apt-get install python
```

Another option is to download the software from the official site[1].

Some systems do not have scientific libraries installed by default. One solution is to install such libraries with commands such as the following:

[1]https://www.python.org/downloads/

```
sudo apt-get install python-numpy python-scipy
sudo apt-get install python-matplotlib python-sklearn
```

Another option is to use the command `pip`, which installs and manages the software packages.

```
sudo apt-get install python-pip
python -m pip install numpy scipy matplotlib sklearn
```

Depending on the operating system, the commands to install might change. Another solution is to install a scientific Python distribution[2]. If you installed Anaconda, a scientific Python distribution, on Windows, then you can install the packages with the software **Anaconda Navigator** or using the command `conda` in the **Anaconda Prompt** as follows.

```
conda install -c anaconda pandas
```

We can use any text editor for coding a program and run it through command line. We can use an integrated development environment (IDE) like Spyder on the operating system or like Jupyter notebook on a web browser. We can run commands interactively in the command-line interface with Python on a notebook or via the command python, an executable file on the operating system command prompt. If the PATH is configured for the operating system to find python, we can run in any directory or folder. Otherwise, we need to find in which directory or folder python was installed. On the Python prompt, we can run the classic `print("Hello World!")`. On Windows, we have an output like the following output.

```
C:\Users\MyLogin\Desktop>..\Anaconda2\python.exe
Python 2.7 |Anaconda, Inc.| (...) [...] on win32
Type "help", "copyright", "credits" or "license" ....
>>> print ("Hello World!")
Hello World!
```

[2]https://scipy.org/install.html

```
>>> exit()

C:\Users\MyLogin\Desktop>
```

We called the Python directly from the directory `..\Anaconda2`. If you installed Anaconda, you can call the **Anaconda Prompt** and can type `python` without worrying about the correct directory. To leave the Python, you can type `quit()`, and to close the Anaconda Prompt, you can type `exit`.

On Linux, things change a little with an output like the following.

```
MyLogin@MyMachine:~$ python
Python 2.7 (default, ...)
[GCC 4.8.4] on linux2
Type "help", "copyright", "credits" or "license" ...
>>> print("Hello World!")
Hello World!
>>> exit()
MyLogin@MyMachine:~$
```

For a quick reference of some Python commands, see Appendix A. Instead of running interactively, we can organize the command in a file and run the file as a script in the command line. In addition, we can use command-line argument to set variables. See the Python programming example in Program 2.1. Line 1 calls the library for collecting the arguments, i.e., the parameters inserted by user. The first argument is collected in lines 4 and the second in line 5. Line 2 calls the library for statistics, which is used in line 8. Notice that the lines 7 and 8 are just one command line. In some situations, we can split the command line. In this case, the signal + in the `print` allows us to split the command line.

Program 2.1
Computing the binomial in Python.

```
1 import sys # load system module
2 import scipy.stats # load statistics module
3
4 f = int(sys.argv[1]) # extract the 1st argument
5 s = int(sys.argv[2]) # extract the 2nd argument
6
7 print ( "Binomial("+str(f)+","+str(s)+") = "+
8 str(scipy.special.comb(f, s, exact=True)) )
```

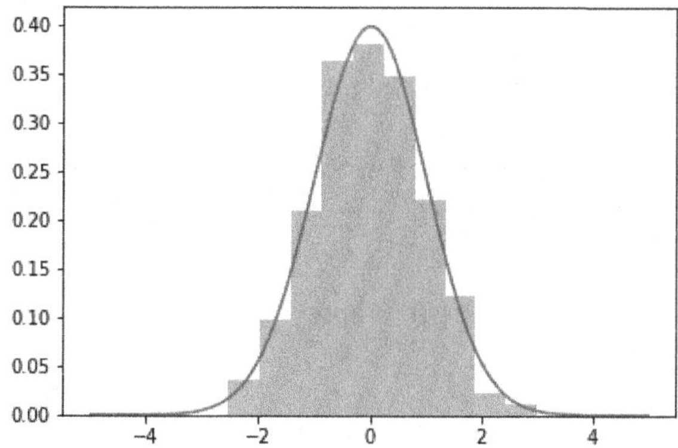

Figure 2.13
Normal distribution in Python.

We can run it in the command line. Depending on the operating system, we can have different prompt in the command line, but we run the Python script as follows.

```
$ python test.py 5 3
Binomial(5,3) = 10

$ python test.py 3 5
Binomial(3,5) = 0
```

Notice that the filename for Python scripts usually has an extension ".py" in lower case.

We can also create a dataset with 500 samples in normal distribution with mean equal to 0 and standard deviation equal to 1. Afterwards, we can plot the histogram and its normal distribution. Figure 2.13 shows one possible result from the script in Program 2.2. This example loads modules just before the usage. Usually, we set all modules at the beginning of the script. Line 2 generates the dataset, which is considered a sample for the example. Line 5 generates the histogram. Line 8 generates the tuple x with 100 elements from -5 to 5. Line 9 computes the probability density function (PDF). Line 12 plots the PDF and line 13 shows the histogram with the plot. With an IDE like Spider, a programmer can explore the variables and learn the Python programming language more effectively.

Program 2.2
Normal distribution in Python

```
1  from scipy.stats import norm
2  sample = norm.rvs(loc=0,scale=1,size=500)
```

```

from pylab import hist
hist(sample, density=True, alpha=.5)

from numpy import linspace
x = linspace(-5,5,100)
pdf = norm.pdf(x)

from pylab import plot,show
plot(x,pdf,'blue')
show()
```

3

Some Common Neural Network Models

Common neural networks are presented in this chapter: multilayer feedforward, associative, Hopfield, recurrent, radial basis function, wavelets, convolutional, and mixture of expert networks. The interface between neural networks and statistical models is mentioned as also is deep learning.

3.1 Multilayer Feedforward Networks

This section deals with networks that use more than one perceptron arranged in layers. This model is able to solve problems that are not linearly separable such as in Figure 3.1 (it is not useful to use ADALINE's layers because the resulting output would be linear). To train, a new learning rule is needed.

The original perceptron learning cannot be extended to the multilayered network, because the output function is not differentiable. If a weight is to be adjusted anywhere in the network, its effect in the output of the network has to be known and hence the error. For this, the derivative of the error or criteria function with respect to that weight must be found, as used in the delta rule for the ADALINE. Hence, a new activation function must be nonlinear, otherwise nonlinearly separable functions could not be implemented, although differentiable. The one that is most often used successfully in multilayered networks is the sigmoid function.

Before we give the general formulation of the generalized delta rule, called backpropagation, we present a numerical illustration for the Exclusive OR (XOR) problem of Figure 3.1(a) with the following inputs and initial weights, using the network of Figure 3.2.

The neural network of Figure 3.2 has one hidden layer with 2 neurons, one input layer with 4 samples and one neuron in the output layer.

Table 3.1 presents the forward computation for one sample $(1, 1)$ in the XOR problem.

Table 3.2 gives the calculus for the backpropagation of the error for the XOR example.

Table 3.3 gives the weight adjustment for this step of the algorithm.

The performance of this update procedure suffers the same drawbacks of any gradient descent algorithm. Several heuristic rules have been suggested to improve its performance, in the neural network literature such as normalizing the inputs to be

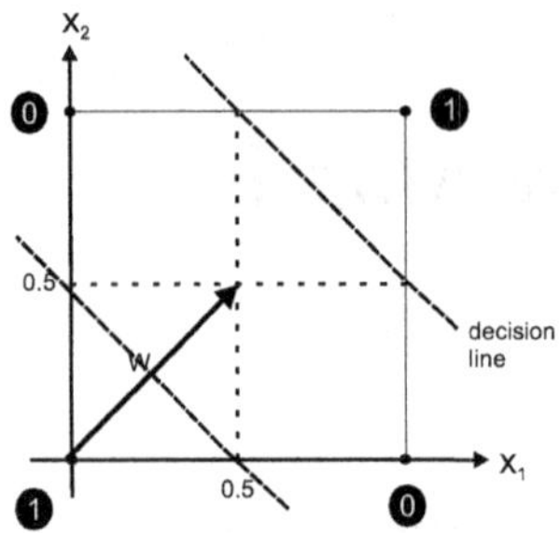

(a) Exclusive OR (XOR).

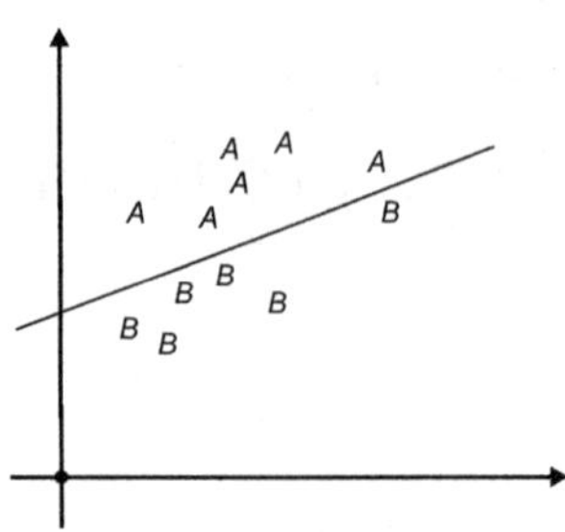

(b) Linear separable class.

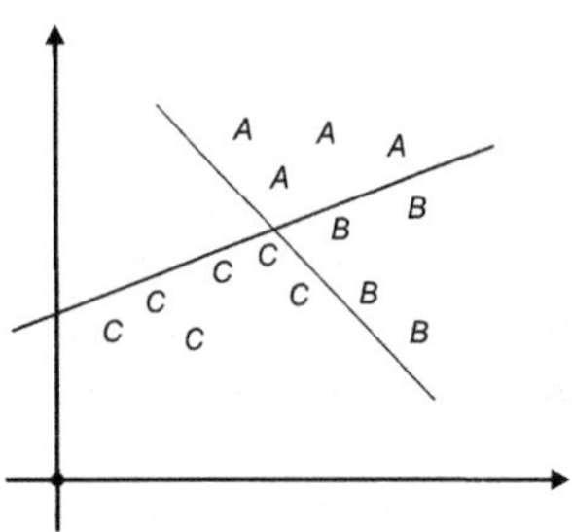

(c) Nonlinear separable.

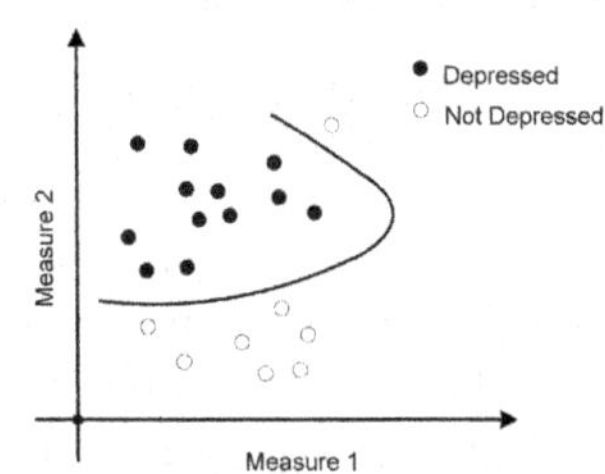

(d) Separable by feedforward (sigmoid) network, psychological measures.

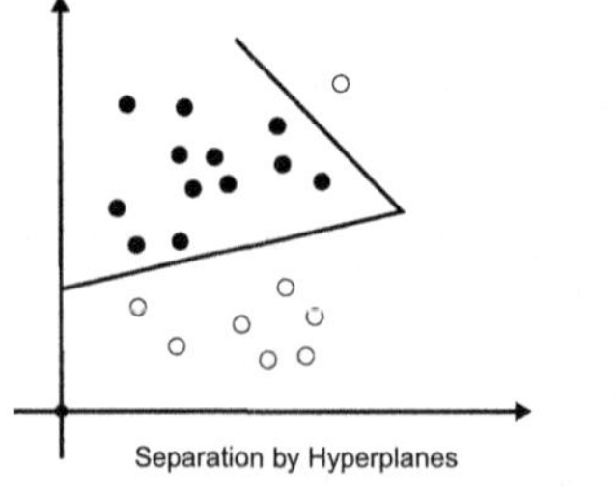

(e) Separable by two-layer perceptron.

Figure 3.1
Classes of problems.

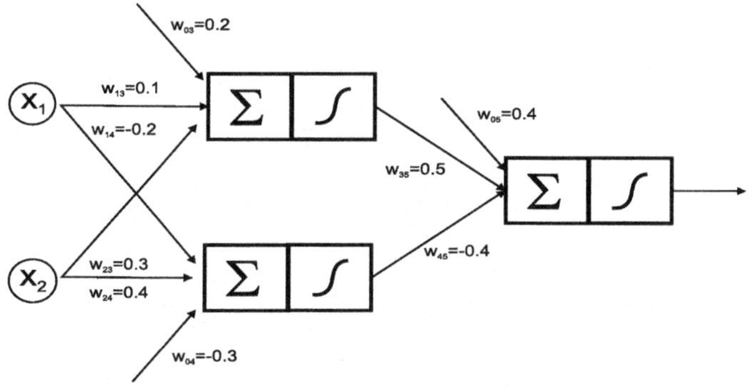

Figure 3.2
Feedforward network.

Table 3.1
Forward computation.

Neuron	Input	Output
3	$.1 + .3 + .2 + .6$	$1/(1 + \exp(-.6)) = .65$
4	$-2.2 + .4 - .3 - .1$	$1/(1 + \exp(.1)) = .48$
5	$.5 \times .65 - .4 \times .48 + .4 = .53$	$1/(1 + \exp(-.53)) = .63$

in the open interval $(0, 1)$. Initial weights are chosen at random in the interval -0.5 and 0.5. The inclusion of a momentum parameter speeds convergence. Here, we only present the general formulation of the algorithm.

The backpropagation algorithm

Let L be the number of layers. The different layers are denoted by $\mathcal{L}_1$ (the input layer), $\mathcal{L}_2, \ldots, \mathcal{L}_M$ (the output layer). We assume the output has M neurons. When neurons i and j are connected a weight ω_{ij} is associated with the connection. In a multilayer feedforward network, only neurons in subsequent layers can be connected:

Table 3.2
Error backpropagation.

Neuron	Error	Tendency Adjustment
5	$.63 \times (1 - .63) \times (0 - .63) = -.147$	$-.147$
4	$.48 \times (1 - .48) \times (-.147) \times (-.4) = .015$	$.015$
3	$.65 \times (1 - .65) \times (-.147) \times (-.4) = -.017$	$-.017$

Table 3.3
Weight adjustment.

Weight	Value
w_{45}	$-.4+(-.147)\times .48=-.47$
w_{35}	$.5+(-.147)\times .65=.40$
w_{24}	$.4+(.015)\times 1=.42$
w_{23}	$.3+(-017)\times 1=.28$
w_{14}	$-.2+(.015)\times 1=.19$
w_{13}	$.1+(-.017)\times 1=.08$
w_{05}	$.4+(-.147)=.25$
w_{04}	$-.3+(.015)=-.29$
w_{03}	$.2+(-.017)=.18$

$$\omega_{ij}=0 \Rightarrow i\epsilon\mathcal{L}_{l+1},\ j\in\mathcal{L}_l\ 1\le l\le L-1.$$

The state is determined by the input and the neuron j and is characterized by z_j. The network operates as follows. The input layer assigns z_j to neurons j in $\mathcal{L}_1$. The output of the neurons in $\mathcal{L}_2$ is

$$z_k=f\left(\sum_{j\in\mathcal{L}_1}\omega_{kj}z_j\right)\ k\in\mathcal{L}_2, \tag{3.1}$$

where f is an activation function that has derivatives for all values of the argument. For a neuron m in an arbitrary layer, we abbreviate

$$a_m=\sum_{j\in\mathcal{L}_{l-1}}\omega_{mj}z_j\ 2\le l\le L, \tag{3.2}$$

where the outputs of layer $l-1$ are known,

$$z_m=f(a_m)\ m\in\mathcal{L}_l. \tag{3.3}$$

We concentrate on the choice of weights when one sample (or pattern) is presented to the network. The desired output, true value, or target will be denoted by $t_k, k\in\mathcal{L}_L\ (k=1,\ldots,M)$.

We want to adapt the initial guess for the weights to decrease

$$D_{LMS}=\sum_{k\in\mathcal{L}_L}(t_k-f(a_k))^2. \tag{3.4}$$

If the function f is nonlinear, then D is a nonlinear function of the weights, the gradient descent algorithm for the multilayer network is as follows.

The weights are updated proportional to the gradient of D with respect to the weights, the weight ω_{mj} is changed by an amount

$$\Delta\omega_{mj} = -\alpha\frac{\partial D}{\partial\omega_{mj}} \quad m \in \mathcal{L}_l,\ j \in \mathcal{L}_{l-1}, \tag{3.5}$$

where α is called the *learning rate*. For the last and next to last layer the updates are given by substituting (3.3) and (3.4) in (3.5), we have

$$\begin{aligned}
\Delta\omega_{mj} &= -\alpha\frac{\partial}{\partial\omega_{mj}}\sum_{p\in\mathcal{L}_L}\left[t_p - f\left(\sum_{q\in\mathcal{L}_{L-1}}\omega_{pq}z_q\right)\right]^2 \\
&= -\alpha 2\sum_{p\in\mathcal{L}_L}(t_p - f(a_p))\,(-f'(a_p))\sum_{g\in\%mcl_{L-1}}\delta_{pm}\delta_{gj}z_g \\
&= -\alpha 2\,(t_m - f(a_m))\,f'(a_m)z_j\ m \in \mathcal{L}_L, j \in \mathcal{L}_{L-1}.
\end{aligned} \tag{3.6}$$

We assume that the neuron is able to calculate the function f and its derivative f'. The error $2\alpha\,(t_m - f(a_m))\,f'(a_m)$ can be sent back (or propagated back) to neuron $j \in \mathcal{L}_{L-1}$. The value z_j is present in neuron j so that the neuron can calculate $\Delta\omega_{mj}$. In an analogous way, it is possible to derive the updates of the weights ending in $\mathcal{L}_{L-1}$, thus

$$\begin{aligned}
\Delta\omega_{mj} &= -\alpha\frac{\partial}{\partial\omega_{mj}}\sum_{p\in\mathcal{L}_L}\left\{t_p - f\left[\sum_{g\in\mathcal{L}_{L-1}}\omega_{pg}f(\omega_{qr}z_r)\right]\right\}^2 \\
&= -\alpha 2\sum_{p\in\mathcal{L}_L}[t_p - f(a_p)]\,(-f'(a_p))\sum_{g\in\%mcl_{L-1}}\delta_{gm}\delta_{xj}z_r \\
&= -\alpha 2\sum_{p\in\mathcal{L}_L}(t_p - f(a_p))\,f'(a_p)\omega_{pm}f'(a_m)z_j, m \in \mathcal{L}_{L-1}, j \in \mathcal{L}_{L-2}.
\end{aligned} \tag{3.7}$$

Now, $2\alpha\,[t_p - f(a_p)]\,f'(a_p)\omega_{pm}$ is the weighted sum of the errors sent from layer $\mathcal{L}_L$ to neuron m in layer $\mathcal{L}_{L-1}$. Neuron m calculates this quantity using $\omega_{pm}, p \in \mathcal{L}_p$, multiplies this quantity $f'(a_m)$, and sends it to neuron j that can calculate $\Delta\omega_{mj}$ and update ω_{mj}.

This weight update is done for each layer in decreasing order of the layers, until $\Delta_{mj}, m \in \mathcal{L}_2, j \in \mathcal{L}_1$. This is the familiar backpropagation algorithm. Note that for the sigmoid (symmetric or logistic) and hyperbolic activation function the properties of its derivatives make it easier to implement the backpropagation algorithm. For further details see Wilde [223] and Picton [166].

3.2 Associative and Hopfield Networks

In unsupervised learning, we have the following groups of learning rules:

- Competitive learning
- Hebbian learning

When the competitive learning is used, the neurons compete among themselves to update the weights. These algorithms have been used in classification, signal extraction, cluster. Examples of networks that use these rules are ART (Adaptive Resonance Theory) and SOM (Self Organized Maps) networks.

Hebbian learning is based on the weight actualization work of neurologist Hebb. This algorithm has been used in characteristics extraction, associative memories. Examples of network that use these rules are: Hopfield and PCA (Principal Component Analysis) networks. The neural networks we have discussed so far have all been trained in a supervised manner. In this section, we start with associative networks, introducing simple rules that allow unsupervised learning. Despite the simplicity of their rules, they form the foundation for the more powerful networks of latter chapters.

The function of an associative memory is to recognize previous learned input vectors, even in the case where some noise is added. We can distinguish three kinds of associative networks:

- *Heteroassociative networks:* map m input vectors $\boldsymbol{x}^1, \boldsymbol{x}^2, \ldots, \boldsymbol{x}^m$ in p-dimensional spaces to m output vector $\boldsymbol{y}^1, \boldsymbol{y}^2, \ldots, \boldsymbol{y}^m$ in k-dimensional space, so that $\boldsymbol{x}^i \to \boldsymbol{y}^i$.
- *Autoassociative networks* are special subsets of heteroassociative networks in which each vector is associated with itself, i.e., $\boldsymbol{y}^i = \boldsymbol{x}^i$ for $i = 1, \ldots, n$. The function of the network is to correct noisy input vector.
- *Pattern recognition* (cluster) networks are a special type of heteroassociative networks. Each vector is associated with the scalar (class) i. The goal of the network is to identify the class of the input pattern.

Figure 3.3 summarizes these three networks.

The three kinds of associative networks can be implemented with a single layer of neurons. Figure 3.4 shows the structure of the heteroassociative network without feedback.

Let $\boldsymbol{W} = (\omega_{ij})$ be the $p \times k$ weight matrix. The row vector $\boldsymbol{x} = (x_1, \ldots, x_p)$ and the identity activation produce the activation

$$\boldsymbol{y}_{1\times k} = \boldsymbol{x}_{1p} \boldsymbol{W}_{p\times k}. \tag{3.8}$$

The basic rule to obtain the elements of W that memorize a set of n associations $(\boldsymbol{x}^i \boldsymbol{y}^i)$ is

$$\boldsymbol{W}_{p\times k} = \sum_{i=1}^{n} \boldsymbol{x}^{i'}_{p\times 1} \circ \boldsymbol{y}^i_{1k}, \tag{3.9}$$

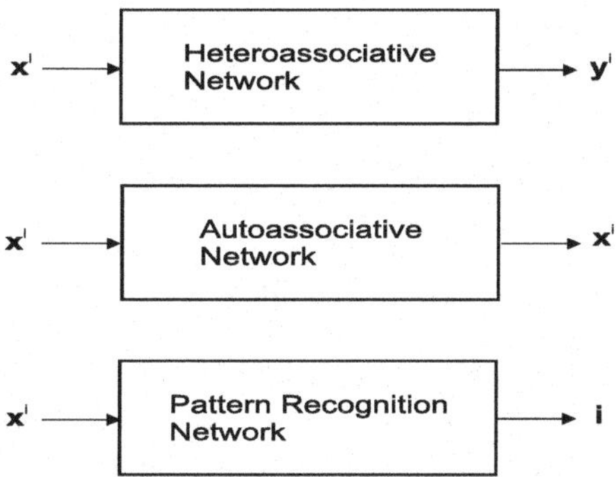

Figure 3.3
Types of associative networks.

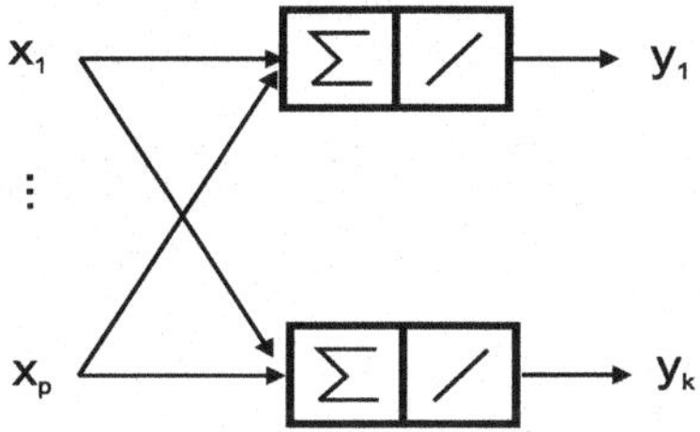

Figure 3.4
Associative network.

that is for n observations. Let X be the $n \times p$ matrix whose rows are the input vector and Y be the $n \times k$ matrix whose rows are the output vector, hence

$$\boldsymbol{W}_{p\times k} = \boldsymbol{X}'_{p\times n} \circ \boldsymbol{Y}_{nk}. \tag{3.10}$$

Therefore, in matrix notation, we also have

$$\boldsymbol{Y} = \boldsymbol{X}\boldsymbol{W}. \tag{3.11}$$

Although without a biological appeal, one solution to obtain the weight matrix is the OLAM - Optimal Linear Associative Memory. If $n = p$, X is square matrix, if it is invertible, the solution of (3.8) is

$$\boldsymbol{W} = \boldsymbol{X}^{-1}\boldsymbol{Y}. \tag{3.12}$$

If $n < p$ and the input vectors are linear independent the optimal solution is

$$\boldsymbol{W}_{p\times k} = \left(\boldsymbol{X}'_{pn}\boldsymbol{X}_{np}\right)^{-1}\boldsymbol{X}'_{pn}\boldsymbol{Y}, \tag{3.13}$$

which minimizes $\sum ||\boldsymbol{W}\boldsymbol{X} - \boldsymbol{Y}||^2$ and $(\boldsymbol{X}'\boldsymbol{X})^{-1}\boldsymbol{X}'$ is the pseudo-inverse of X. If the input vectors are linear dependent, we have to use regularization theory (similar to ridge regression).

The biological appeal to update the weight is to use the Hebb rule

$$\boldsymbol{W} = \eta \boldsymbol{X}'\boldsymbol{Y} \tag{3.14}$$

as each sample presentation where η is a learning parameter in $(0, 1)$.

The association networks are useful for recognition (such as face, figure, etc.), compression (such as images), and function optimization (such as Hopfield network).

We will present some of these association networks using some examples of the literature.

Consider the example Abdi et al. [1] where we want to store a set of schematic faces given in Figure 3.5 in an associative memory using Hebbian learning. Each face is represented by a 4-dimensional binary vector, denoted by $\boldsymbol{x}^i$ in which a given element corresponds to one of the features from Figure 3.6.

Figure 3.5
Schematic faces for Hebbian learning. Each face is made of four features (hair, eyes, nose, and mouth), taking the value $+1$ or -1.

FEATURE
+1
-1
CELL
Hair
1
Eyes
2
Nose
3
Mouth
4

Figure 3.6
The features used to build the faces in Figure 3.5.

Suppose that we start with $\eta = 1$.

$$
\boldsymbol{W}_0 = \begin{bmatrix} 0 & 0 & 0 & 0 \\ 0 & 0 & 0 & 0 \\ 0 & 0 & 0 & 0 \\ 0 & 0 & 0 & 0 \end{bmatrix} \tag{3.15}
$$

The first face is stored in the memory by modifying the weights to

$$
\boldsymbol{W}_1 = \boldsymbol{W}_0 + \begin{bmatrix} 1 \\ 1 \\ 1 \\ -1 \end{bmatrix} \begin{bmatrix} 1 & 1 & 1 & -1 \end{bmatrix} = \begin{bmatrix} 1 & 1 & 1 & 1 \\ 1 & 1 & 1 & 1 \\ 1 & 1 & 1 & 1 \\ -1 & -1 & -1 & -1 \end{bmatrix}, \tag{3.16}
$$

the second face is stored by

$$
\boldsymbol{W}_2 = \boldsymbol{W}_1 + \begin{bmatrix} -1 \\ -1 \\ 1 \\ 1 \end{bmatrix} \begin{bmatrix} -1 & -1 & 1 & 1 \end{bmatrix} = \begin{bmatrix} 2 & 2 & 0 & 0 \\ 2 & 2 & 0 & 0 \\ 0 & 0 & 2 & -2 \\ 0 & 0 & -2 & +2 \end{bmatrix}, \tag{3.17}
$$

and so on until

$$
\boldsymbol{W}_{10} = \boldsymbol{W}_9 + \begin{bmatrix} 1 \\ 1 \\ -1 \\ 1 \end{bmatrix} \begin{bmatrix} 1 & 1 & -1 & 1 \end{bmatrix} = \begin{bmatrix} 10 & 2 & 2 & 6 \\ 2 & -10 & 2 & -2 \\ 2 & 2 & 10 & -2 \\ 6 & -2 & -2 & 10 \end{bmatrix}. \tag{3.18}
$$

Another important association network is the Hopfield network. In its presentation, we follow closely Chester [38]. Hopfield network is recurrent and fully interconnected with each neuron feeding its output in all others. The concept was that all the neurons would transmit signals back and forth to each other in a closed feedback loop until their state became stable. The Hopfield topology is illustrated in Figure 3.7.

Figure 3.8 provides examples of a Hopfield network with nine neurons. The operation is as follows:

Assume that there are three 9-dimensional vectors α, β and γ.

$$
\alpha = (101010101) \ \beta = (110011001) \ \gamma = (010010010). \tag{3.19}
$$

The substitution of -1 for each 0 in these vectors transforms them into bipolar form $\alpha^*, \beta^*, \gamma^*$.

$$
\begin{aligned}
\alpha^* &= (1-11-11-11-11)' \\
\beta^* &= (11-1-111-1-11)' \\
\gamma^* &= (-11-1-11-1-11-1)'.
\end{aligned} \tag{3.20}
$$

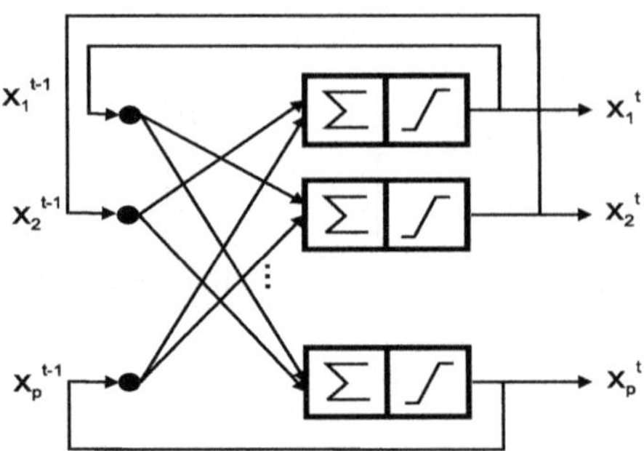

Figure 3.7
Hopfield network architecture.

The weight matrix is

$$\boldsymbol{W} = \alpha^* \alpha^{*\prime} + \beta^* \beta^{*\prime} + \gamma^* \gamma^{*\prime}, \tag{3.21}$$

when the main diagonal is zeroed (no neuron feeding into itself). This becomes the matrix in Figure 3.8 of the example.

The study of the dynamics of the Hopfield network and its continuous extension is an important subject, which includes the study of network capacity. How many neurons are needed to store a certain number of patterns? Note that the bipolar form allows greater storage energy (Lyapunov) function of the system. These functions are more related to optimization than to statistics applications.

We end this section with an example from Fausett [69] on comparison of data using a BAM (Bidirectional Associative Memory) network. The BAM network is used to associate letters with simple bipolar codes.

Consider the possibility of using a (discrete) BAM network (with bipolar vectors) to map two simple letters (given by 5×3 patterns) to the following bipolar codes given by (3.22) as follows.

$$\begin{array}{ccc} \cdot & \# & \cdot \\ \# & \cdot & \# \\ \# & \# & \# \\ \# & \cdot & \# \\ \# & \cdot & \# \\ \multicolumn{3}{c}{(-1,1)} \end{array} \qquad \begin{array}{ccc} \cdot & \# & \# \\ \# & \cdot & \cdot \\ \# & \cdot & \cdot \\ \# & \cdot & \cdot \\ \cdot & \# & \# \\ \multicolumn{3}{c}{(1,1)} \end{array} \tag{3.22}$$

The weight matrices are given by (3.23) as follows.

	A	B	C	D	E	F	G	H	I
From A	0	-1	1	-1	1	1	1	-3	3
B	-1	0	-3	-1	1	1	-3	1	-1
C	1	-3	0	1	-1	-1	3	-1	1
D	-1	-1	1	0	-3	1	1	1	-1
E	1	1	-1	-3	0	-1	-1	-1	1
F	1	1	-1	1	-1	0	-1	-1	1
G	1	-3	3	1	-1	-1	0	-1	1
H	-3	1	-1	1	-1	-1	-1	0	-3
I	3	-1	1	-1	1	1	1	-3	0

Synaptic weight matrix, nodes $A - I$.

Example 1			Example 2		Example 3		
Start	a	b	Start	a	Start	a	b
$A = 1$	$A = 1$	$A = 1$	$A = 1$	$A = 1$	$A = 1$	$A = 1$	$A = 1$
$B = 1$	$B \to 0$	$B = 0$	$B = 1$	$B = 1$	$B = 0$	$B = 0$	$B = 0$
$C = 1$	$C = 1$	$C = 1$	$C = 0$	$C = 0$	$C = 0$	$C \to 1$	$C = 1$
$D = 1$	$D \to 0$	$D \to 1$	$D = 0$	$D = 0$	$D = 1$	$D = 1$	$D = 1$
$E = 1$	$E \to 0$	$E = 0$	$E = 1$	$E = 1$	$E = 0$	$E = 0$	$E = 0$
$F = 1$	$F \to 0$	$F \to 1$	$F = 1$	$F = 1$	$F = 0$	$F \to 1$	$F = 1$
$G = 1$	$G = 1$	$G = 1$	$G = 0$	$G = 0$	$G = 1$	$G = 1$	$G = 1$
$H = 1$	$H \to 0$	$H = 0$	$H = 0$	$H = 0$	$H = 0$	$H = 0$	$H = 0$
$I = 1$	$I = 1$	$I = 1$	$I = 0$	$I \to 1$	$I = 0$	$I \to 1$	$I = 1$

Figure 3.8
Example of Hopfield network. The matrix shows the synaptic weights between neurons. Each example shows an initial state for the neurons (representing an input vector). State transitions are shown by arrows, as individual nodes are updated one at a time in response to the changing states of the other nodes. In example 2, the network output converges to a stored vector. In examples 1 and 3, the stable output is a spontaneous attractor, identical to none of the three stored vectors. The density of 1-bits in the stored vectors is relatively high and the vectors share enough common bit locations to put them far from orthogonality, all of which may make the matrix somewhat quirky, perhaps, contributing to its convergence toward the spontaneous attractor in examples 1 and 3.

$$\begin{array}{ccc} \text{(to store } A \to -11) & (C \to 11) & (\boldsymbol{W}, \text{ to store both}) \\ \begin{bmatrix} 1 & -1 \\ -1 & 1 \\ 1 & -1 \\ -1 & 1 \\ 1 & -1 \\ -1 & 1 \\ -1 & 1 \\ -1 & 1 \\ -1 & 1 \\ -1 & 1 \\ 1 & -1 \\ -1 & 1 \\ -1 & 1 \\ 1 & -1 \\ -1 & 1 \end{bmatrix} & \begin{bmatrix} -1 & -1 \\ 1 & 1 \\ 1 & 1 \\ 1 & 1 \\ -1 & -1 \\ -1 & -1 \\ 1 & 1 \\ -1 & -1 \\ -1 & -1 \\ 1 & 1 \\ -1 & -1 \\ -1 & -1 \\ -1 & -1 \\ 1 & 1 \\ 1 & 1 \end{bmatrix} & \begin{bmatrix} 0 & -2 \\ 0 & 2 \\ 2 & 0 \\ 0 & 2 \\ 0 & -2 \\ -2 & 0 \\ 0 & 2 \\ -2 & 0 \\ -2 & 0 \\ 0 & 2 \\ 0 & -2 \\ -2 & 0 \\ -2 & 0 \\ 2 & 0 \\ 0 & 2 \end{bmatrix} \end{array} . \tag{3.23}$$

To illustrate the use of a BAM, we first demonstrate that the net gives the correct Y vector when presented with the $\boldsymbol{x}$ vector for either pattern A or the pattern C:

INPUT PATTERN A

$$\begin{bmatrix}-11 & -11 & -111111 & -111 & -11\end{bmatrix} \boldsymbol{W} = (-14, 16) \to (-1, 1). \tag{3.24}$$

INPUT PATTERN C

$$\begin{bmatrix}-1111 & -1 & -11 & -1 & -11 & -1 & -1 & -11\end{bmatrix} \boldsymbol{W} = (14, 16) \to (1, 1). \tag{3.25}$$

To see the bidirectional nature of the net, observe that the Y vectors can also be used as input. For signals sent from the Y-layer to the X-layer, the weight matrix is the transpose of the matrix $\boldsymbol{W}$, i.e.,

$$\boldsymbol{W}^T = \begin{bmatrix} 0 & 0 & 2 & 0 & 0 & -2 & 0 & -2 & -2 & 0 & 0 & -2 & -2 & 2 & 0 \\ 0 & 0 & 2 & 0 & 0 & -2 & 0 & -2 & -2 & 0 & 0 & -2 & -2 & 2 & 0 \end{bmatrix}. \tag{3.26}$$

For the input vector associated with pattern A, namely, $(-1, 1)$, we have

$$\begin{aligned} &(-1,1)\boldsymbol{W}^T = \\ &(-1,1)\begin{bmatrix} 0 & 0 & 2 & 0 & 0 & -2 & 0 & -2 & -2 & 0 & 0 & -2 & -2 & 2 & 0 \\ 0 & 0 & 2 & 0 & 0 & -2 & 0 & -2 & -2 & 0 & 0 & -2 & -2 & 2 & 0 \end{bmatrix} \\ &= \begin{bmatrix}-2 & 2 & -2 & 2 & -2 & 2 & 2 & 2 & 2 & 2 & -2 & 2 & 2 & -2 & 2\end{bmatrix} \\ &\to \begin{bmatrix}-1 & 1 & -1 & 1 & -1 & 1 & 1 & 1 & 1 & 1 & -1 & 1 & 1 & -1 & 1\end{bmatrix}, \end{aligned} \tag{3.27}$$

which is the pattern A.

Similarly, if we input the vector associated with pattern C, namely, $(1, 1)$, we obtain

$$\begin{aligned} &(-1,1)\boldsymbol{W}^T = \\ &(-1,1)\begin{bmatrix} 0 & 0 & 2 & 0 & 0 & -2 & 0 & -2 & -2 & 0 & 0 & -2 & -2 & 2 & 0 \\ -2 & 2 & 0 & 2 & -2 & 0 & & 0 & 0 & 2 & 2 & 0 & 0 & 0 & 2 \end{bmatrix} \\ &= \begin{bmatrix}-2 & 2 & 2 & 2 & -2 & -2 & 2 & -2 & -2 & 2 & -2 & -2 & -2 & -2 & 2\end{bmatrix} \\ &\to \begin{bmatrix}-1 & 1 & 1 & 1 & -1 & -1 & 1 & -1 & -1 & 1 & -1 & -1 & -1 & 1 & 1\end{bmatrix}, \end{aligned} \tag{3.28}$$

which is the pattern C.

The net can also be used with noisy input for the x vector, the y vector, or both.

3.3 Radial Basis Function Networks

The radial basis function is a single hidden layer feedforward network with a linear output transfer function and a nonlinear transfer function $h(\cdot)$ in the hidden layer. Many types of nonlinearity may be used.

The most general formula for the radial function is

$$h(x) = \Phi\left((x-c)R^{-1}(x-c)\right), \tag{3.29}$$

where Φ is the function used (Gaussian, multiquadratic, etc.), c is the center and R the metric. The common functions are: Gaussian $\Phi(z) = e^{-z}$, multiquadratic $\Phi(z) = (1+z)^{1/2}$, inverse multiquadratic $\Phi(z) = (1+z)^{-1/2}$ and the Cauchy $\Phi(z) = (1+z)^{-1}$.

If the metric is Euclidean $R = r^2 I$ for some scalar radius and the equation simplifies to

$$h(x) = \Phi\left(\frac{(x-c)^T(x-c)}{r^2}\right). \tag{3.30}$$

The essence of the difference between the operation of radial basis function and multilayer perceptron is shown in Figure 3.1(e) for a hypothetical classification of psychiatric patients, with classification by alternative networks: multilayer perceptron (left) and radial basis functions networks (right).

Multilayer perceptron separates the data by hyperplanes while the radial basis function clusters the data into a finite number of ellipsoid regions.

A simplification is the 1-dimensional input space in which case we have

$$h(x) = \Phi\left(\frac{(x-c)^2}{\sigma^2}\right). \tag{3.31}$$

Figure 3.9 illustrates the alternative forms of $h(x)$ for $c = 0$ and $r = 1$.

To see how the radial basis function network (RBFN) works, we consider some examples in Wu and McLarty [225]. Consider two measures (from questionnaires) of mood to diagnose depression. The algorithm that implemented the radial basis function application determines that seven cluster sites were needed for this problem. A constant variability term, $\sigma^2 = 1$, was used for each hidden unit. The Gaussian function was used for activation of the hidden layers. Figure 3.10 shows the network.

Suppose we have the scores $x' = (4.8; 1.4)$. The Euclidean distance from this vector and the cluster mean of the first hidden unit $\mu_1' = (0.36; 0.52)$ is

$$D = \sqrt{(4.8-.36)^2 + (.14-.52)^2} = .398. \tag{3.32}$$

The output of the first hidden layer is

$$h_1(D^2) = e^{-.398^2/2\times(.1)} = .453. \tag{3.33}$$

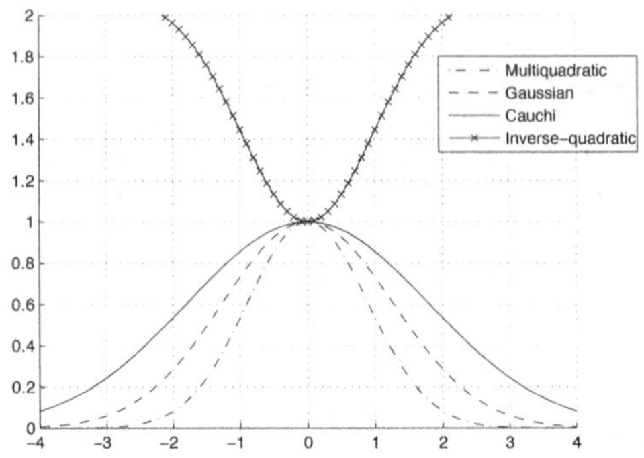

Figure 3.9
Radial basis functions $h(.)$.

This calculation is repeated for each remaining hidden unit. The final output is

$$(.453)(2.64) + (.670)(-.64) + (543)(5.25) + \ldots + \\ +(-1)(0.0092) = 0.0544. \tag{3.34}$$

Since the output is close to 0, the patient is classified as not depressed.

The general form of the output is $y = f(x) = \sum_{i=1}^{m} w_i h_i(x)$, that is the function f is expressed as a linear combination of m fixed functions (called basis function in analogy to the concept of a vector being composed as a combination of basis vectors).

Training radial basis function networks proceeds in two steps. First the hidden layer parameters are determined as a function of the input data and then the weights in between the hidden and output layer are determined from the output of the hidden layer and the target data.

Therefore, we have a combination of unsupervised training to find the cluster and the cluster parameters and in the second step a supervised training that is linear in the parameters. Usually in the first step the k-means algorithm of clustering is used, but other algorithms can also be used including SOM and ART (chapter 4).

3.4 Wavelet Neural Networks

The approximation of a function in terms of a set of orthogonal basis functions is familiar in many areas, in particular in statistics. An example is the Fourier expansion, where the basis consists of sines and cosines of differing frequencies. Another is the Walsh expansion of categorical time sequence or square wave, where the concept of

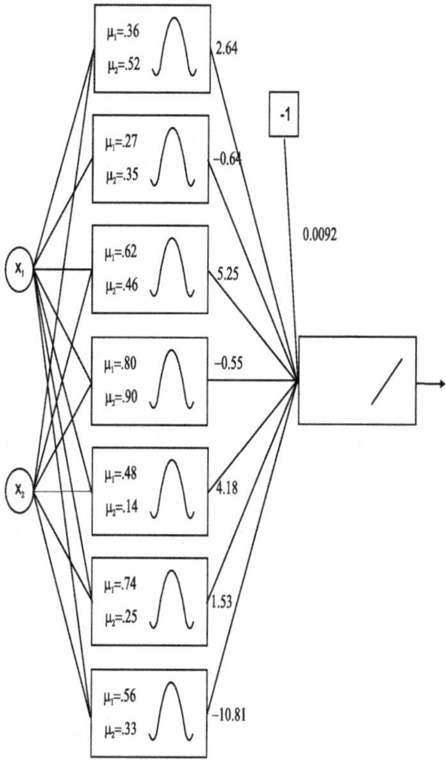

Figure 3.10
Radial basis network.

frequency is replaced by that of sequence, which gives the number of “zero crossings" of the unit interval. In recent years, there has been considerable interest in the development and use of an alternative type of basis function: *wavelets*.

The main advantage of wavelets over Fourier expansion is that wavelets operate as localized approximators in comparison to the global characteristic of the Fourier expansion. Therefore, wavelets have significant advantages when the data is non-stationary.

In this section, we will outline why and how neural networks have been used to implement wavelet applications. Some wavelets results are first presented. See also Morettin [146] and Hubbard [91].

3.4.1 Wavelets

To present the wavelet basis of functions, we start with a *father wavelet* or *scaling* function ϕ such that

$$\phi(x) = \sqrt{2}\sum_k l_k \phi(2x-k), \tag{3.35}$$

usually, normalized as $\int_{-\infty}^{\infty} \phi(x)dx = 1$. A mother wavelet is obtained through

$$\psi(x) = \sqrt{2}\sum h_k \phi(2x-k) \tag{3.36}$$

where l_k and h_k are related through

$$h_k = (-1)^k l_k. \tag{3.37}$$

The equations (3.35) and (3.36) are called *dilatation equations*, the coefficients l_k, h_k are low-pass and high-pass filters, respectively.

We assume that these functions generate an orthonormal system of $L_2(\mathcal{R})$, denoted $\{\phi_{j_0,k}(x)\} \cup \{\psi_{jk}(x)\}_{j\geq j_0,k}$ with $\phi_{j_0,k}(x) = 2^{j_0/2}\phi(2^{j_0}x-k),\ \psi_{j,k}(x) = 2^{j/2}\psi(2^j x-k)$ for $j \geq j_0$ on the coarse scale.

For any $f \in L_2(\mathcal{R})$, we may consider the expansion

$$f(x) = \sum_{k=-\infty}^{\infty} \alpha_k \phi_{j_0,k}(x) + \sum_{j\geq j_0}\sum_{k=-\infty}^{\infty} \beta_{j,k}\psi_{j,k}(x) \tag{3.38}$$

for some coarse scale j_0 and where the true wavelet coefficients are given by

$$\alpha_k = \int_{-\infty}^{\infty} f(x)\phi_{j_0,k}(x)dx \ \ \beta_{j,k} = \int_{-\infty}^{\infty} f(x)\psi_{j,k}(x)dx. \tag{3.39}$$

An estimate will take the form

$$\hat{f}(x) = \sum_{k=-\infty}^{\infty} \hat{\alpha}_k \phi_{j_0,k}(x) + \sum_{j\geq j_0}\sum_{k=-\infty}^{\infty} \hat{\beta}_{j,k}\psi(x), \tag{3.40}$$

where $\hat{\alpha}_k$ and $\hat{\beta}_k$ are estimates of α_k and β_{jk}, respectively.

Several issues are of interest in the process of obtaining the estimate $\hat{f}(x)$:

(i) the choice of the wavelet basis;

(ii) the choice of thresholding policy (which $\hat{\alpha}_j$ and $\hat{\beta}_{jk}$ should enter in the expression of $\hat{f}(x)$; the value of j_0; for finite sample, the number of parameters in the approximation);

(iii) the choice of further parameters appearing in the thresholding scheme;

(iv) the estimation of the scale parameter (noise level).

These issues are discussed in Morettin [147] and Abramovich et al. [2].

Concerning the choice of wavelet basis, some possibilities include the Haar wavelet, which is useful for categorical-type data. It is based on

$$\phi(x) = 1 \quad 0 \leq x < 1$$
$$\psi(x) = \begin{cases} 1 & 0 \leq x < 1/2 \\ -1 & 1/2 \leq x < 1 \end{cases} \tag{3.41}$$

and the expansion is then

$$f(x) = \alpha_0 + \sum_{j=0}^{J}, \sum_{k=0}^{2^J-1} \beta_{jk}\psi_{jk}(x). \tag{3.42}$$

Other common wavelets are: the Shannon with mother function

$$\psi(x) = \frac{\sin(\pi x/2)}{\pi x/2}\cos(3\pi x/2), \tag{3.43}$$

and the Mortet wavelet with mother function

$$\psi(x) = e^{iwx}e^{-x^2/2}, \tag{3.44}$$

where $i = \sqrt{-1}$ and w is a fixed frequency. Figure 3.11 shows the plots of these wavelets. Figure 3.12 illustrates the localized characteristic of wavelets.

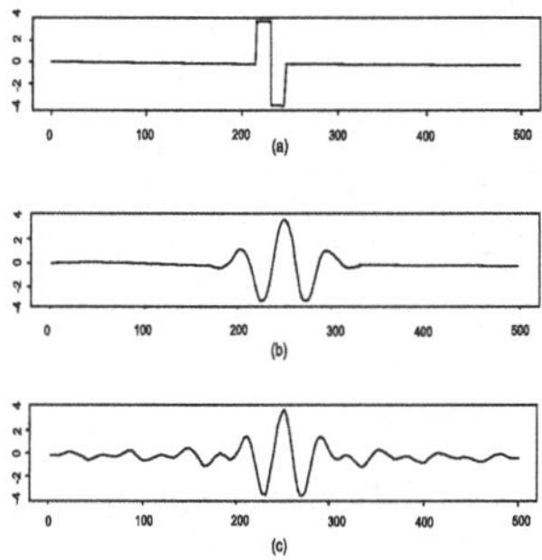

Figure 3.11
Some wavelets: (a) Haar (b) Morlet (c) Shannon.

Figure 3.12 gives an example of a wavelet transform.

In the figure 3.12, a wavelet (b) is compared successively to different sections of a function (a). The product of the section and the wavelet is a new function; the area delimited by that function is the wavelet coefficient. In (c), the wavelet is compared to a section of the function that looks like the wavelet. The product of the two is always positive, giving the big coefficient shown in (d). (The product of two negative

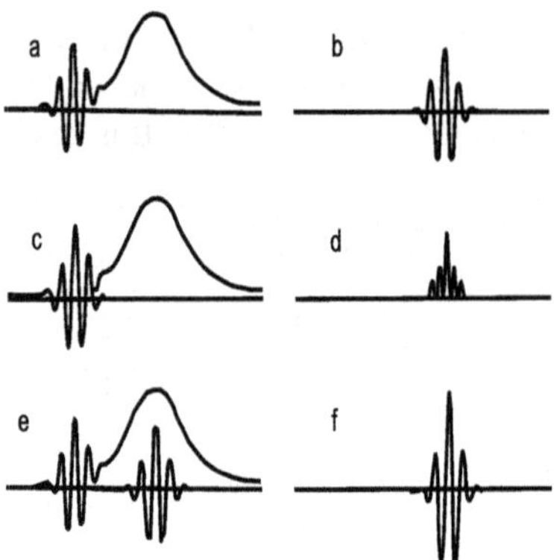

Figure 3.12
Wavelet operation.

functions is positive.) In (e) the wavelet is compared to a slowly changing section of the function, giving the small coefficients shown in (f). The signal is analyzed at different scales, using wavelets of different widths. "You play with the width of the wavelet in order to catch the rhythm of the signal," says Yves Meyer.

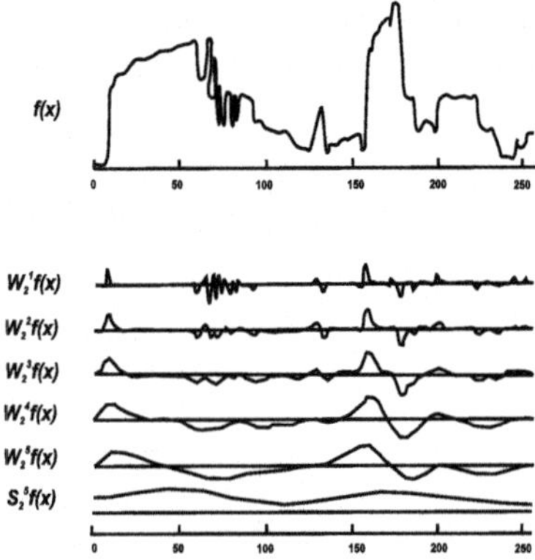

Figure 3.13
Example of a wavelet transform.

Figure 3.13 (top) shows the original signal, followed by the wavelet transforms of the signal over five scales differing by a factor of 2. The bottom graph depicts the remaining low frequencies.

3.4.2 Wavelet Networks and Radial Basis Wavelet Networks.

On the one hand, we have seen (Section 2.7 - Kolmogorov Theorem) that neural networks have been established as a general approximation tool for fitting nonlinear models from input and output data. On the other hand, wavelets decomposition emerges as powerful tool for approximation. Because of the similarity between the discrete wavelet transform and a one-hidden-layer neural network, the idea of combining both wavelets and neural networks has been proposed. This has resulted in the wavelet network discussed in Iyengar et al. [99] and summarized here.

There are two main approaches to form a wavelet network. In the first, the wavelet component is decoupled from the estimation components of the perceptron. In essence the data is decomposed on some wavelets and this is fed into the neural network. We call this approach the radial basis wavelet neural networks, which is shown in Figure 3.14.

In the second approach, the wavelet theory and neural networks are combined into a single method. In wavelet networks, both the position and dilatation of the wavelets as well as the weights are optimized. The neuron of a wavelet network is a multidimensional wavelet in which the dilatation and translation coefficients are considered as neuron parameters. The output of a wavelet network is a linear combination of several multidimensional wavelets.

Figures 3.15 and 3.16 give a general representation of a wavelet neuron (wavelon) and a wavelet network.

Observation. It is useful to rewrite the wavelet basis as

$$\Psi_{a,b} = |a|^{-1}\left(\frac{x-b}{a}\right) \quad a > 0, -\infty < b < \infty \text{ with } a = 2^j, b = k2^{-j}.$$

A detailed discussion of these networks is given in Zhang and Benveniste [231]. See also Iyengar et al. [99] and Martin and Morris [141] for further references.

3.5 Mixture-of-Experts Networks

Consider the problem of learning a function in which the form of the function varies in different regions of the input space (see also radial basis and wavelets networks sections). These types of problems can be made easier using mixture of experts or modular networks.

In these types of networks, we assigned different expert networks to table each of the different regions, and then used an extra "gating" network, which also sees the input data, to deduce which of the experts should be used to determine the output. A more general approach would allow the discovery of a suitable decomposition as part of the learning process.

Figure 3.17 shows several possible decompositions.

These types of decompositions are related to mixture of models, generalized additive and tree models in statistics (see Section 5.2).

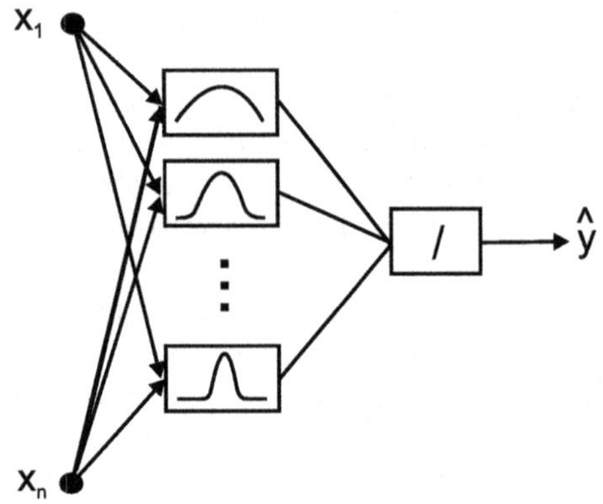

(a) Function approximation.

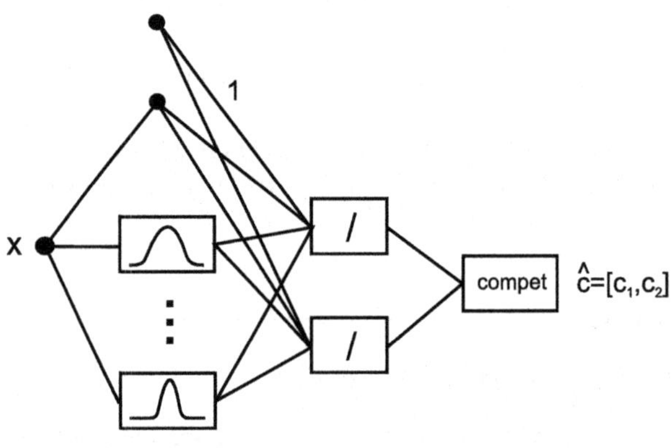

(b) Classification.

Figure 3.14
Radial basis wavelet network.

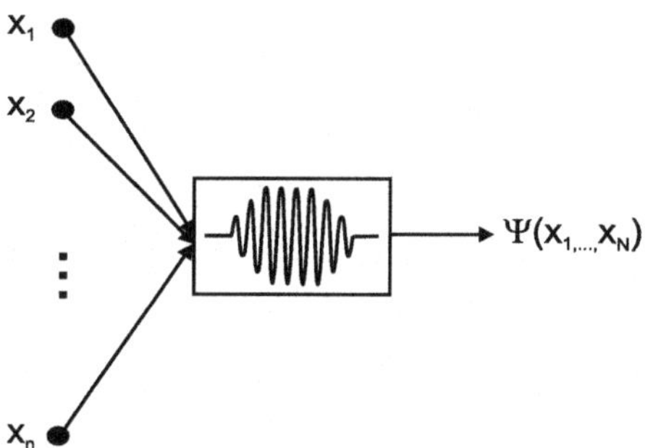

Figure 3.15
Wavelon.

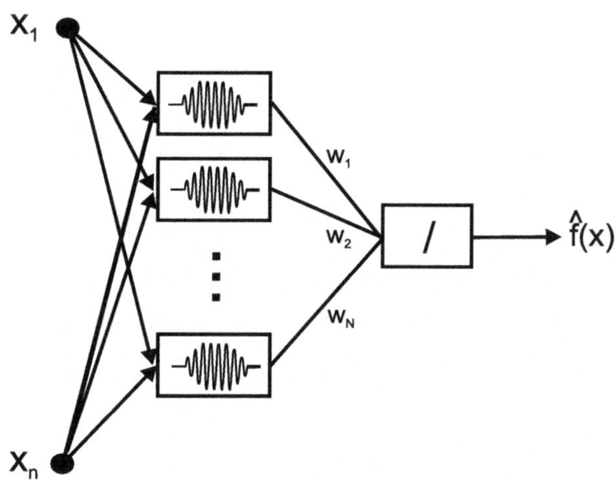

Figure 3.16
Wavelet network - $f(x) = \sum_{i=1}^{N} w_i \Psi\left(\frac{x_i - a_i}{b_i}\right)$.

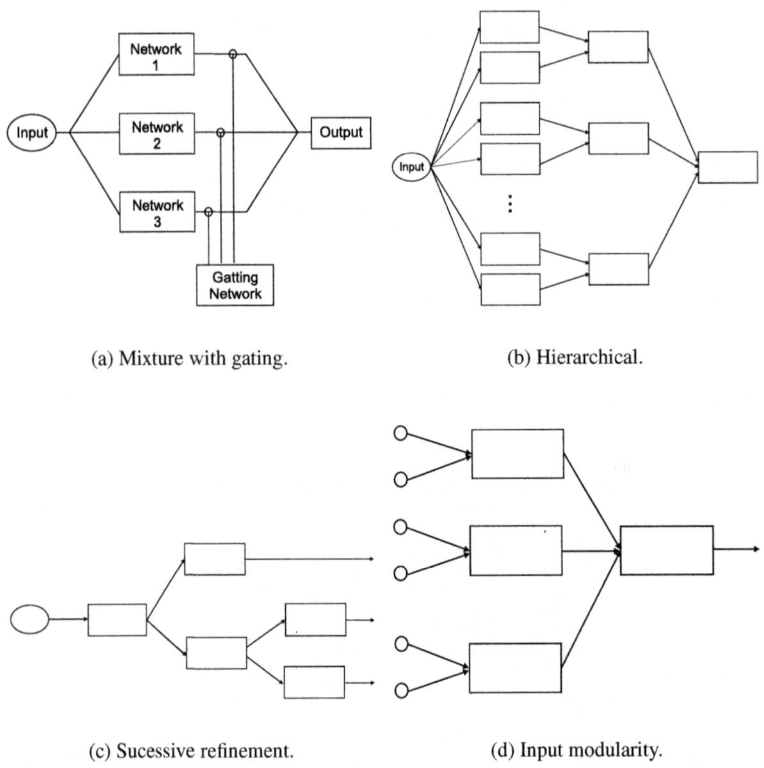

(a) Mixture with gating.

(b) Hierarchical.

(c) Sucessive refinement.

(d) Input modularity.

Figure 3.17
Mixture-of-experts networks.

Further details and references can be seen in Mehrotra et al. [142] and Bishop [20]. Applications can be seen in Jordan and Jacobs [102], Ohno-Machado et al. [155], Peng et al. [163], Desai et al. [58], Cheng [37], Ciampi and Lechevalier [43], and Ng and McLachlan [151].

3.6 Neural Network and Statistical Model Interfaces

The remaining chapters of this book apply these networks to different problems of data analysis in place of statistical models. It is therefore convenient to review the statistical bibliography relating neural network and statistical models and statistical problems in general.

The relationship of multilayer feedforward network with several linear models is presented by Arminger and Polozik [10]. They show that these networks can perform equivalently to: linear regression, linear discrimination, multivariate regression, probit regression, logit regression, generalized linear models, generalized additive models with known smoothing functions, projection pursuit, and LISREL (with mixed dependent variables). Other related papers are: Cheng and Titterington [36], Sarle [187], Stern [199], Warner and Misra [216], Schumacher et al. [189], Vach et al. [213], and De Veaux et al. [55].

Attempts to compare performance of neural networks were made extensively but no conclusive result seems to be possible; some examples are Balakrishnan et al. [13], Mangiameli et al. [138], Mingoti and Lima [144] for cluster analysis and Tu [211], Ennis et al. [63], and Schwarzer et al. [190] for logistic regression.

Statistical intervals for neural networks were studied in Tibshirani [207], Hwang and Ding [92], Chryssolouriz et al. [40] and De Veaux et al. [55]. For Bayesian analysis for neural networks, see Mackay [136], Neal [149], Lee, Lee [122, 119], and Insua and Müller [97], Paige and Butler [158], and the review of Titterington [208].

Theoretical statistical results are presented in Murtagh [148], Poli and Jones [168], Jordan [101], Faraggi and Simon [65], Ripley [177], Lowe [134], Martin and Morris [141], the review papers by Cheng and Titterington [36] and Titterington [208]. Asymptotics and further results can be seen in a series of papers by White [219, 220, 221, 222], Swanson and White [203, 204], and Kuan and White [113]. Books relating neural networks and statistics are: Smith [195], Bishop [20], Ripley [177], Neal [149] and Lee [119]. For a practical view in engineering and applied science, see Silva et al. [193]. Neural networks are applied also in cryptography (see Coutinho et al. [51]) and in smart grids (see Raza and Khosravi [175]), for more information about smart grids see Borges de Oliveira [23].

Pitfalls on the use of neural networks versus statistical modeling are discussed in Tu [211], Schwarzer et al. [190] and Zhang [230]. Further references on the interfaces: statistical tests, methods (e.g., survival, multivariate, etc) and others are given in the corresponding chapters of the book.

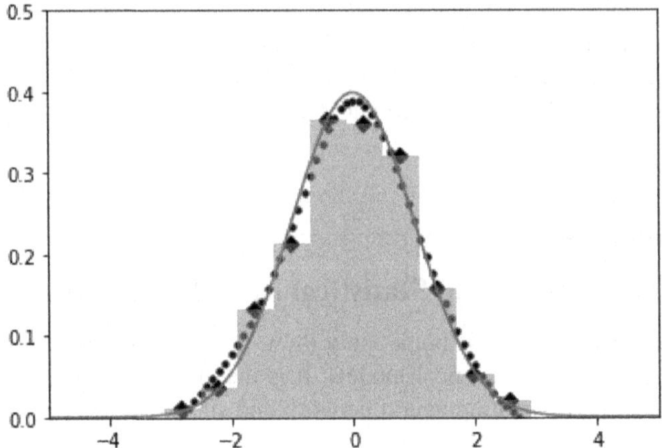

Figure 3.18
Normal distribution in Python with a neural network.

3.7 Some Common Neural Networks in Python

In this section, neural networks are used to fit data, to compute classification, and to memorize using a Hopfield network and BAM.

3.7.1 Fitting Data

In 2.2, we learned how to create a dataset, to plot its histogram and to compute its PDF. Someone might ask how to compute a normal distribution using a neural network. Indeed, it is not recommended to use a neural network when we know the function or algorithm that solves the problem. However, a neural network might provide a faster solution than known algorithms for big data, i.e., when we have too much data. To approximate a normal distribution with a neural network, we use the command `MLPRegressor` from the module `neural_network` in the library Scikit-learn `sklearn`, see Pedregosa et al. [162]. The code is in the Program 3.1, which uses the same command from the example in Program 2.2 and produces the result in Figure 3.18. However, we rewrote it in a more professional way. Lines from 1 to 4 call the modules where each `as` provides an abbreviation for the modules. Excluding the neural network model, both programs use the same modules. In this way, we need to write the abbreviation before using each command from the module. It might look strange, but it can avoid conflict when more than one module has commands with the same name.

Program 3.1
Normal distribution in Python with a neural network.

```
1  import scipy.stats as ss
2  import numpy as np
3  import matplotlib.pyplot as plt
4  import sklearn.neural_network as nn
5
6  sample = ss.norm.rvs(loc=0,scale=1,size=500)
7  H=plt.hist(sample,density=True,alpha=.5,color=['#999999'])
8  x = np.linspace(-5,5,100)
9  pdf = ss.norm.pdf(x)
10 plt.plot(x,pdf,'b')
11
12 y=H[0]
13 x = np.frombuffer(H[1][0:10]+0.25).reshape(-1, 1)
14 plt.scatter(x, y, c='b', marker="D")
15
16 NN = nn.MLPRegressor(activation='tanh', solver='lbfgs')
17 NN.fit(x, y)
18
19 test_x = np.arange(-5, 5, 0.1).reshape(-1, 1)
20 test_y = NN.predict(test_x)
21
22 plt.scatter(test_x,test_y, s=10, c='r', marker="o")
23
24 plt.xlim(-5, 5)
25 plt.ylim(0, 0.5)
26
27 plt.show()
```

In Program 3.1, we defined the activation function `tanh` in line 16. Other options are `identity`, `logistic`, and `relu`. If we do not set the parameter `activation='tanh'`, the default is currently `relu`, which is more interesting when the neural network has several hidden layers. In this case, the network architectures are also known as deep learning or deep neural networks. The default is one hidden layer with 100 neurons. Thus, the command

```
MLPRegressor(activation='tanh')
```

is equivalent to

```
MLPRegressor(hidden_layer_sizes=(100), activation='tanh').
```

If we want to set a neural network with four hidden layers containing 500, 400, 300, 200, 100 neurons, respectively, we need to set the command parameter as `hidden_layer_sizes=(500, 400, 300, 200, 100)`.

In addition to using the neural network to fit a curve through the points, we can also use the command `MLPRegressor` to approximate a curve. In Program 3.2, we find a neural network to approximate the curve $\sin(x) + x$. Figure 3.19 shows its output.

Program 3.2
Approximation of $\sin(x) + x$ with n neural network.

```
import numpy as np
import matplotlib.pyplot as plt
from sklearn.neural_network import MLPRegressor

x = np.arange(0, 10, .005).reshape(-1, 1)
y = (np.sin(x)+x).ravel()

NN = MLPRegressor(activation='tanh', solver='lbfgs')
NN.fit(x, y)

test_x = np.arange(0, 10, 0.5).reshape(-1, 1)
test_y = NN.predict(test_x)

plt.scatter(x, y, c='b', marker=".")
plt.scatter(test_x, test_y, s=25, c='r', marker="D")

plt.xlim(0, 10)
plt.ylim(0, 10)

plt.show()
```

In Program 3.2, if we replace $\sin(x) + x$ for $\sin(2x) + x$, the difficult for fitting the curve increases. Therefore, we need to enlarge the hidden layers.

Another parameter that we can change to test is the solver. We are using `lbfgs`, a quasi-Newton method. Other options are the stochastic gradient descent `sgd` and another optimizer based on stochastic gradient `adam`.

3.7.2 Classification

The command to create a neural network for classification is `MLPClassifier`. Luckily, its parameters are similar to the parameters used in the command `MLPRegressor`.

Program 3.3 shows an example for classification. The libraries are loaded from lines 1 to 5. Lines 7 and 8 create the dataset that we call `sample` with 8 points split into two types. Lines 10 and 13 adjust the axis. Line 15 creates a neural network with two hidden layers, which is trained in line 16. If we insert the parameter `verbose` in line 15, line 16 will show information about the training. We can insert a parameter

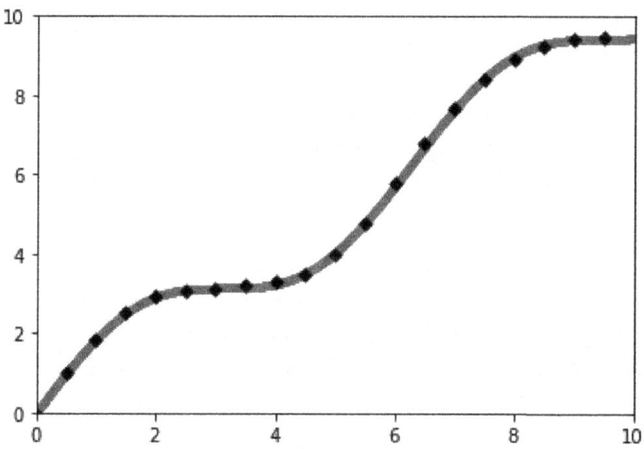

Figure 3.19
Approximation of $\sin(x) + x$ with a neural network.

such as:

```
MLPClassifier(hidden_layer_sizes=(100,100), verbose=True)
```

Program 3.3
Classification with a neural network.

```
import numpy as np
from matplotlib import pyplot as plt
from matplotlib.colors import ListedColormap as cm
from sklearn import datasets as db, \
                     neural_network as nn \

sample, types = db.make_moons(
          n_samples=8, random_state=0)

x_min, x_max = sample[:,0].min()-1, sample[:,0].max()+1
y_min, y_max = sample[:,1].min()-1, sample[:,1].max()+1
X, Y = np.meshgrid(np.arange(x_min, x_max, .1),
                   np.arange(y_min, y_max, .1))

NN = nn.MLPClassifier(hidden_layer_sizes=(100,100))
NN.fit(sample, types)

Z = NN.predict_proba(np.c_[X.ravel(), Y.ravel()])[:, 1]
```

```
19 Z = Z.reshape(X.shape)
20 plt.contourf(X, Y, Z, levels=[0,.5])
21
22 plt.scatter(sample[:, 0], sample[:, 1],
23                  c=types, cmap=cm(['#FF0000', '#0000FF'])
                        ↪ )
24
25 plt.show()
```

Lines from 18 to 20 delimit the classification as shown in Figure 3.20, which presents the output of the Program 3.3 without `verbose=True`. The output changes, when we run the example. The command in lines 22 and 23 shows the two types of points in the sample.

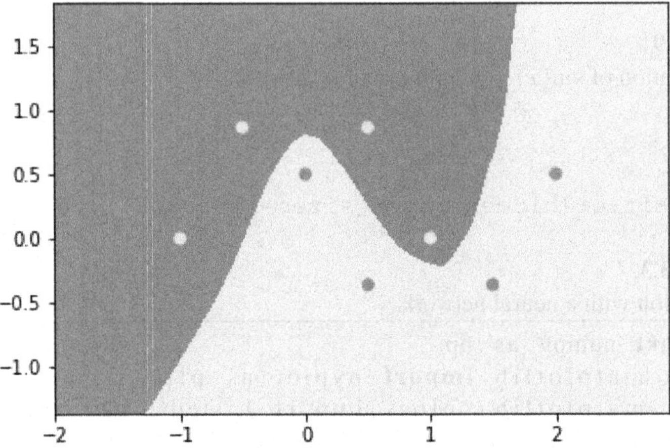

Figure 3.20
Classification with a neural network.

Considering that the limits of the classification change when we run the program, someone may ask what the accuracy would be. To evaluate the accuracy, let us create a dataset with 20 points split into two types. Instead of training the network with all points, we can split again into points for training and points for testing the limits of the classification. This approach is known as classification and cross validation. Program 3.4 shows this example, which is very similar to Program 3.3.

Program 3.4
Classification and cross validation with a neural network.

```
import numpy as np
from matplotlib import pyplot as plt
from matplotlib.colors import ListedColormap as cm
```

```
from sklearn import datasets, model_selection, neural_network
sample, types = datasets.make_moons( noise=.5,
        n_samples=20, random_state=0)
sample_train,sample_test,types_train,types_test \
        =model_selection.train_test_split(
        sample, types, test_size=10, shuffle=False)
x_min, x_max = sample[:,0].min()-1, sample[:,0].max()+1
y_min, y_max = sample[:,1].min()-1, sample[:,1].max()+1
X, Y = np.meshgrid(np.arange(x_min, x_max, .1),
                   np.arange(y_min, y_max, .1))
NN = neural_network.MLPClassifier(hidden_layer_sizes=(100,100))
NN.fit(sample_train, types_train)
trein_accuracy= NN.score(sample_train,types_train)
print("\nTrein Accuracy = %5.2f" %trein_accuracy)
test_accuracy = NN.score(sample_test,types_test)
print("Test  Accuracy = %5.2f" %test_accuracy)
Z = NN.predict_proba(np.c_[X.ravel(), Y.ravel()])[:, 1]
Z = Z.reshape(X.shape)
plt.contourf(X, Y, Z, levels=[0,.5])
plt.scatter(sample_train[:, 0], sample_train[:, 1],
            c=types_train, cmap=cm(['#FF0000', '#0000FF']))
plt.scatter(sample_test[:, 0], sample_test[:, 1], marker='x',
            c=types_test, cmap=cm(['#FF0000', '#0000FF']))
```

```
41 plt.show()
```

To create a more interesting example, we add noise in the `sample` with the parameter `noise=.5` in line 6. Lines from 23 to 26 compute the accuracy for training and test data. The former should be 1 if the network has enough neurons. However, the noise might decrease the accuracy for the test data. In the output presented in Figure 3.21, the train accuracy is 1 and the test accuracy is 0.9.

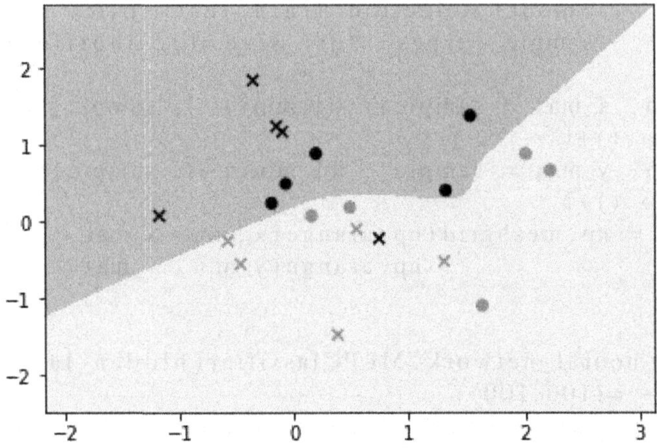

Figure 3.21
Classification and cross validation with a neural network.

3.7.3 Hopfield Networks

Let us use another library, NeuPy – Neural Networks in Python. Depending on the versions of the libraries installed on your operating system, you might need to install an earlier version of Neupy[1]. We can select a specific version with the command `pip`, as in the example bellow.

```
python -m pip install neupy==0.6.5
```

You may find the list of versions in the repository Python Package Index (PyPI)[2].

Instead of using -1 and 1 to set the data, this library uses zeros and ones for Hopfield networks. We draw the symbols of the elementary operations with matrix

[1] http://neupy.com
[2] https://pypi.org/project/neupy/#history

in lines from 10 to 40 in Program 3.5. A function named `image` that receives the matrices and outputs the images is presented in lines from 5 to 8.

Program 3.5

A Hopfield network that memorizes the symbols of operations.

```
1  import numpy as np
2  import matplotlib.pyplot as plt
3  from neupy import algorithms
4
5  def image(matrix):
6      plt.imshow(matrix.reshape(5,5), cmap=plt.cm.
         ↪ binary);
7      plt.axis('off')
8      plt.show()
9
10 plus = np.matrix([
11     0, 0, 1, 0, 0,
12     0, 0, 1, 0, 0,
13     1, 1, 1, 1, 1,
14     0, 0, 1, 0, 0,
15     0, 0, 1, 0, 0,
16 ])
17
18 times = np.matrix([
19     1, 0, 0, 0, 1,
20     0, 1, 0, 1, 0,
21     0, 0, 1, 0, 0,
22     0, 1, 0, 1, 0,
23     1, 0, 0, 0, 1,
24 ])
25
26 minus = np.matrix([
27     0, 0, 0, 0, 0,
28     0, 0, 0, 0, 0,
29     1, 1, 1, 1, 1,
30     0, 0, 0, 0, 0,
31     0, 0, 0, 0, 0,
32 ])
33
34 obelus = np.matrix([
35     0, 0, 1, 0, 0,
36     0, 0, 0, 0, 0,
37     1, 1, 1, 1, 1,
38     0, 0, 0, 0, 0,
39     0, 0, 1, 0, 0,
```

```
40 ])
41
42 image(plus)
43 image(times)
44 image(minus)
45 image(obelus)
46
47 operations = np.concatenate([plus, times, minus,
       obelus])
48 nn = algorithms.DiscreteHopfieldNetwork()
49 nn.train(operations)
50
51 part = np.matrix([
52       0, 0, 0, 0, 1,
53       0, 0, 0, 0, 0,
54       0, 0, 0, 0, 0,
55       0, 1, 0, 0, 0,
56       1, 0, 0, 0, 1,
57 ])
58 image(part)
59
60 result = nn.predict(part)
61 image(result)
```

The Program 3.5 outputs the image of for elementary operations in lines from 42 to 45, joins the symbols in line 47, memorizes the symbols in lines 48 and 49, shows part of a symbol in line 58, predicts in line 60, and presents the correct symbol in line 61. Figure 3.22 depicts first the operations, namely, plus 3.22(a), times 3.22(b), minus 3.22(c), obelus 3.22(d). Afterwards, it presents the symbol with part of its information in Figure 3.22(e). Finally, it outputs the correct result in Figure 3.22(f).

To use BAM, we can change Program 3.5 to run the Program 3.6, which does the same, but instead of outputting an image, it outputs a vector associated to a symbol in lines from 36 to 39. In line 43, we set the BAM. The output is

```
[[0 1 0 0]]
```

in agreement with line 37.

Program 3.6
A BAM network that memorizes the symbols of operations.

```
1 import numpy as np
2 from neupy import algorithms
3
```

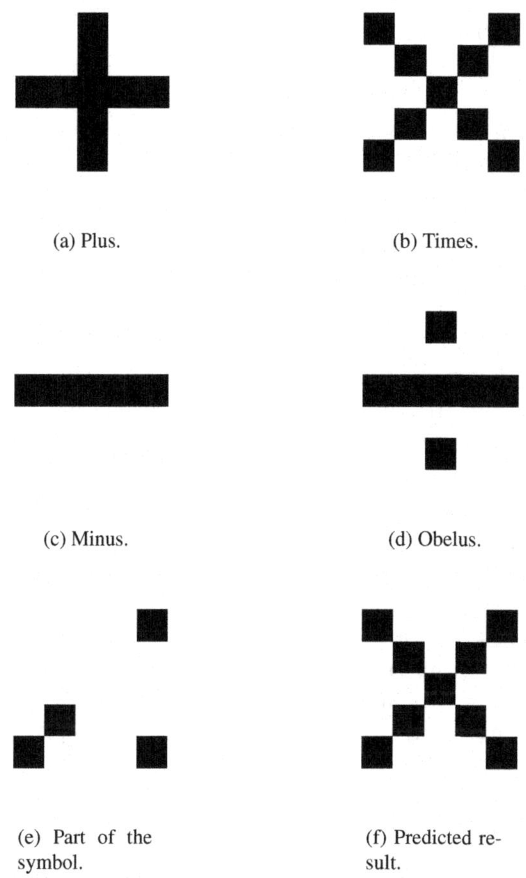

(a) Plus. (b) Times.

(c) Minus. (d) Obelus.

(e) Part of the symbol. (f) Predicted result.

Figure 3.22
The output of Program 3.5.

```
plus = np.matrix([
        0, 0, 1, 0, 0,
        0, 0, 1, 0, 0,
        1, 1, 1, 1, 1,
        0, 0, 1, 0, 0,
        0, 0, 1, 0, 0,
])

times = np.matrix([
        1, 0, 0, 0, 1,
        0, 1, 0, 1, 0,
        0, 0, 1, 0, 0,
```

```
    0, 1, 0, 1, 0,
    1, 0, 0, 0, 1,
])

minus = np.matrix([
    0, 0, 0, 0, 0,
    0, 0, 0, 0, 0,
    1, 1, 1, 1, 1,
    0, 0, 0, 0, 0,
    0, 0, 0, 0, 0,
])

obelus = np.matrix([
    0, 0, 1, 0, 0,
    0, 0, 0, 0, 0,
    1, 1, 1, 1, 1,
    0, 0, 0, 0, 0,
    0, 0, 1, 0, 0,
])

plus_hint = np.matrix([[1, 0, 0, 0]])
times_hint = np.matrix([[0, 1, 0, 0]])
minus_hint = np.matrix([[0, 0, 1, 0]])
obelus_hint = np.matrix([[0, 0, 0, 1]])

operations = np.concatenate([plus, times, minus,
    ↪ obelus])
hints = np.concatenate([plus_hint, times_hint,
    ↪ minus_hint, obelus_hint])
nn = algorithms.DiscreteBAM()
nn.train(operations, hints)

part = np.matrix([
    0, 0, 0, 0, 1,
    0, 0, 0, 0, 0,
    0, 0, 0, 0, 0,
    0, 1, 0, 0, 0,
    1, 0, 0, 0, 1,
])

part, result = nn.predict(part)
print(result)
```

4

Multivariate Statistics Neural Network Models

In this chapter, neural networks are used to perform multivariate statistical analysis, for instance, cluster and scaling network analysis, competitive, learning vector quantization, adaptive resonance theory (ART) networks and self-organizing map (SOM) networks. In addition, this chapter presents dimensional reduction methods: linear and nonlinear principal component analysis (PCA), independent component analysis (ICA), factor analysis (FA), correspondence analysis (CA), multidimensional scaling. Moreover, the networks for these methods are presented: PCA networks, nonlinear PCA networks, FA networks, CA networks, and ICA networks.

4.1 Cluster and Scaling Networks

4.1.1 Competitive Networks

In this section, we present some networks based on competitive learning used for clustering and a network with a similar role to that of multidimensional scaling (Kohonen map network).

The most extreme form of a competition is the Winner Take All and is suited to cases of competitive unsupervised learning.

Several networks discussed in this section use the same learning algorithm (weight updating).

Assume that there is a single layer with M neurons and that each neuron has its own set of weights w_j. An input x' is applied to all neurons. The activation function is the identity. Figure 4.1 illustrates the architecture of the Winner Take All network.

The node with the best response to the applied input vector is declared the winner, according to the winner criterion

$$O_l = \min_{j=1\ldots c}\left\{d^2(y_j, w_j)\right\} \tag{4.1}$$

where d is a distance measure of y_j from the cluster prototype or center. The change of weights is calculated according to

$$w_i(k+1) = w_l(k) + \alpha(x^i - w_i(k)) \tag{4.2}$$

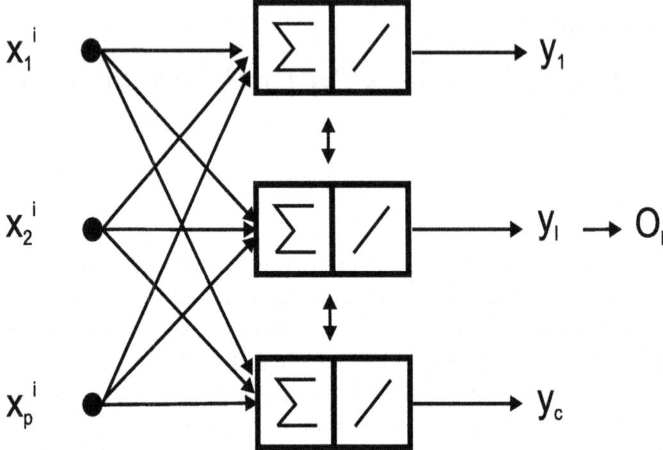

Figure 4.1
General Winner Take All network.

where α is the learning rate and may be decreasing at each iteration.

We now present an example due to Mehrotra et al. [142] to illustrate this simple competitive network.

Let the set n consists of 6 three-dimensional vectors $x^1, x^2, \ldots, x^6$,

$$\begin{aligned} n = \{x_1^1 = (1.1, 1.7, 1.8), x_2^2 = (0, 0, 0), x_3^3 = (0, 0.5, 1.5), \\ x_4^4 = (1, 0, 0), x_5^5 = (0.5, 0.5, 0.5), x_6^6 = (1, 1, 1)\}. \end{aligned} \tag{4.3}$$

We begin a network containing three input nodes and three processing units (output nodes) A, B, C. The connection strengths of A, B, C are initially chosen randomly and are given by the weight matrix

$$W(0) = \begin{pmatrix} w_1 \\ w_2 \\ w_3 \end{pmatrix} = \begin{pmatrix} 0.2 & 0.7 & 0.3 \\ 0.1 & 0.1 & 0.9 \\ 1 & 1 & 1 \end{pmatrix}. \tag{4.4}$$

To simplify computations, we use a learning rate $\eta = 0.5$ and update weights (Equation (4.2)). We compare squared Euclidean distances to select the winner: $d_{j.l}^2 \equiv d^2(w_j, i_l)$ refers to the squared Euclidean distance between the current position of the processing node j from the lth pattern.

For $t = 1$, the sample presented is $x_1^1 = (1.1, 1.7, 1.8)$. The squared Euclidean distance between A and x_1^1 is $d_{1.1}^2 = (1.1 - 0.2)^2 + (1.7 - 0.7)^2 + (1.8 - 0.3)^2 = 4.1$. Similarly, $d_{2.1}^2 = 4.4$ and $d_{3.1}^2 = 1.1$.

C is the "winner" since $d_{3.1}^2 < d_{1.1}^2$ and $d_{3.1}^2 < d_{2.1}^2$. A and B are therefore not perturbed by this sample whereas C moves halfway towards the sample (since $\eta = 0.5$). The resulting weight matrix is

$$W(1) = \begin{pmatrix} w_1 \\ w_2 \\ w_3 \end{pmatrix} = \begin{pmatrix} 0.2 & 0.7 & 0.3 \\ 0.1 & 0.1 & 0.9 \\ 1.05 & 1.35 & 1.4 \end{pmatrix}. \tag{4.5}$$

$t = 2$: Sample presented $x_2^2 = (0, 0, 0), d_{1.2}^2 = 0.6, d_{2.2}^2 = 0.8, d_{3.2}^2 = 4.9$, hence A is the winner. The weights of A are updated. The resulting modified weight vector is $w_1 : (0.1, 0.35, 0.15)$.

Similarly, we have:

$t = 3$: Sample presented $x_3^3 = (0, 0.5, 1.5), d_{2.3}^2 = 0.5$ is the least, hence B is the winner and it is updated. The resulting modified weight vector is w_2 : $(0.05, 0.3, 1.2)$.

$t = 4$: Sample presented $x_4^4 = (1, 0, 0), d_{1.4}^2 = 1, d_{2.4}^2 = 2.4, d_{3.4}^2 = 3.8$, hence A is the winner and it is updated: $w_1 : (0.55, 0.2, 0.1)$.

$t = 5$: $x_5^5 = (0.5, 0.5, 0.5)$ is presented, winner A is updated: $w_1(5) = (0.5, 0.35, 0.3)$.

$t = 6$: $x_6^6 = (1, 1, 1)$ is presented, winner C is updated: $w_3(6) = (1, 1.2, 1.2)$.

$t = 7$: x_1^1 is presented, winner C is updated: $w_3(7) = (1.05, 1.45, 1.5)$.

$t = 8$: Sample presented x_2^2. Winner A is updated to $w_1(8) = (0.25, 0.2, 0.15)$.

$t = 9$: Sample presented x_3^3. Winner B is updated to $w_1(9) = (0, 0.4, 1.35)$.

$t = 10$: Sample presented x_4^4. Winner A is updated to $w_1(10) = (0.6, 0.1, 0.1)$.

$t = 11$: Sample presented x_5^5. Winner A is updated to $w_1(11) = (0.55, 0.3, 0.3)$.

$t = 12$: Sample presented x_6^6. Winner C is updated and $w_3(12) = (1, 1.2, 1.25)$.

At this stage, the weight matrix is

$$W(12) = \begin{pmatrix} w_1 \\ w_2 \\ w_3 \end{pmatrix} = \begin{pmatrix} 0.55 & 0.3 & 0.3 \\ 0 & 0.4 & 1.35 \\ 1 & 1.2 & 1.25 \end{pmatrix}. \tag{4.6}$$

The node A becomes repeatedly activated by the samples x_2^2, x_4^4, and x_5^5, the node B by x_3^3 alone, and the node C by x_1^1 and x_6^6. The centroid of x_2^2, x_4^4, and i_5 is $(0.5, 0.2, 0.2)$, and convergence of the weight vector for node A towards this location is indicated by the progression

$$\begin{aligned} &(0.2, 0.7, 0.3) \to (0.1, 0.35, 0.15) \to (0.55, 0.2, 0.1) \to (0.5, 0.35, 0.3) \to \\ &(0.25, 0.2, 0.15) \to (0.6, 0.1, 0.1) \to (0.55, 0.3, 0.3) \ldots \end{aligned} \tag{4.7}$$

It is attractive to interpret the competitive learning algorithm as a clustering process, but the merits of doing so are debatable, as illustrated by a simple example where the procedure clusters the numbers 0, 2, 3, 5, 6 in groups $A = (0, 2), (3, 5), (6)$ instead of the more natural grouping $A = (0), B = (2, 3)$ and $C = (5, 6)$.

For further discussion on the relation of this procedure to k-means clustering see Mehrotra et al. [142, p. 168].

4.1.2 Learning Vector Quantization (LVQ)

Here again we present the results from Mehrotra et al. [142, p. 173].

Unsupervised learning and clustering can be useful preprocessing steps for solving classification problems. A learning vector quantizer (LVQ) is an application of winner-take-all networks for such tasks and illustrates how an unsupervised learning mechanism can be adapted to solve supervised learning tasks in which class membership is known for every training pattern.

Each node in an LVQ is associated with an arbitrarily chosen class label. The number of nodes chosen for each class is roughly proportional to the number of training patterns that belong to that class, making the assumption that each cluster has roughly the same number of patterns. The new updating rule may be paraphrased as follows.

When pattern i from class $C(i)$ is presented to the network, let the winner node j^* belong to class $C(j^*)$. The winner j^* is moved towards the pattern i if $C(i) = C(j^*)$ and away from i otherwise.

This algorithm is referred to as LVQ1, to distinguish it from more recent variants of the algorithm. In the LVQ1 algorithm, the weight update rule uses a learning rate $\eta(t)$ that is a function of time t, such as $\eta(t) = 1/t$ or $\eta(t) = a[1 - (t/A)]$ where a and A are positive constants and $A > 1$.

The following example illustrates the result of running the LVQ1 algorithm on the input samples of the previous section.

The data are cast into a classification problem by arbitrarily associating the first and last samples with Class 1, and the remaining samples with Class 2. Thus, the training set is: $T = \{(x_1, 1), (x_2, 0), \ldots, (x_6, 1)\}$ where $x^1 = (1.1, 1.7, 1, 8), x^2 = (0, 0, 0), x^3 = (0, 0.5, 1.5), x^4 = (1, 0, 0), x^5 = (0.5, 0.5, 0.5)$, and $x^6 = (1, 1, 1)$.

The initial weight matrix is

$$W(0) = \begin{pmatrix} w_1: & 0.2 & 0.7 & 0.3 \\ w_2: & 0.1 & 0.1 & 1.9 \\ w_3: & 1 & 1 & 1 \end{pmatrix}. \tag{4.8}$$

Since there are twice as many samples in Class 2 as in Class 1, we label the first node (w_1) as associated with Class 1, and the other two nodes with Class 2. Let $\eta(t) = 0.5$ until $t = 6$, then $\eta(t) = 0.25$ until $t = 12$, and $\eta(t) = 0.1$ thereafter. Only the change in a single weight vector in each weight updates iteration of the network, instead of writing out the entire weight matrix.

1. Sample x_1^1, winner w_3 (distance 1.07), w_3 changed to (0.95, 0.65, 0.60).
2. Sample x_2^2, winner w_1 (distance 0.79), w_1 changed to (0.30, 1.05, 0.45).
3. Sample x_3^3, winner w_2 (distance 0.73), w_2 changed to (0.05, 0.30, 1.20).
4. Sample x_4, winner w_3 (distance 0.89), w_3 changed to (0.97, 0.33, 0.30).

... ...

156. Sample x_6, winner w_1 (distance 0.56), w_1 changed to (1.04, 1.33, 1.38).
157. Sample x_1, winner w_1 (distance 0.57), w_1 changed to (1.05, 1.37, 1.42).
158. Sample x_2, winner w_3 (distance 0.58), w_3 changed to (0.46, 0.17, 0.17).

159. Sample x_3, winner w_2 (distance 0.02), w_2 changed to (0.00, 0.49, 1.48).
160. Sample x_4, winner w_3 (distance 0.58), w_3 changed to (0.52, 0.15, 0.15).
161. Sample x_5, winner w_3 (distance 0.50), w_3 changed to (0.52, 0.18, 0.18).
162. Sample x_6, winner w_1 (distance 0.56), w_1 changed to (1.05, 1.33, 1.38).

Note that associations between input samples and weight vectors stabilize by the second cycle of pattern presentations, although the weight vectors continue to change.

Mehrotra et al. [142] also mentioned a variation called the LVQ2 learning algorithm with a different learning rule used instead and an update rule similar to that of LVQ1.

In one application of the procedure, Patazopoulos et al. [160] applied a LVQ network to discriminate benign from malignant cells on the basis of the extracted morphometric and textural features. Another LVQ network was also applied in an attempt to discriminate at the patient level. The data consisted of 470 samples of voided urine from an equal number of patients with urothelial lesions. For the purpose of the study 45,452 cells were measured. The training sample used 30% of the patients and cells, respectively. The study included 50 cases of lithiasis, 61 cases of inflammation, 99 cases of benign prostatic hyperplasia, 5 cases of in situ carcinoma, 71 cases of grade I transitional cell carcinoma of the bladder (TCCB) and 184 cases of grade II and grade III TCCB.

The application enables the correct classification of 95.42% of the benign cells and 86.75% of the malignant cells. At the patient level the LVQ network enables the correct classification of 100% of benign cases and 95.6% of the malignant cases. The overall accuracy rates were 90.63% and 97.57%, respectively. Maksimovic and Popovic [137] compared alternative networks to classify arm movements in tetraplegics. The following hand movements were studied: I - up/down proximal to the body on the lateral side; II - left/right above the level of the shoulder; III - internal/external rotation of the upper arm (humerus).

Figure 4.2 presents the correct classification percentages for the neural networks used.

Vuckovic et al. [215] used neural networks to classify alert versus drowsy states from 1-second long sequences of full-spectrum EEGs in arbitrary subjects. The experimental data was collected on 17 subjects. Two experts in the EEG interpretation provided the expertise to train the three artificial neural networks (ANNs) used in the comparison: one-layer perceptron with identity activation (Widrow-Hoff optimization), perceptron with sigmoid activation (Levenberg-Marquart optimization), and LVQ network.

The LVQ network yielded best classification. For validation 12 volunteers were used and the matching between the human assessment and the network was $94.37 \pm 1.95\%$ using the t statistics.

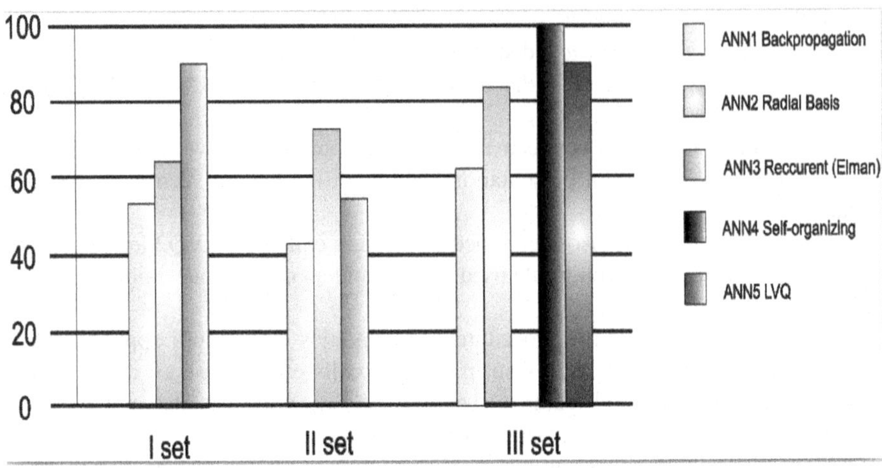

Figure 4.2
Success of classification for all ANNs and all tested movement trials.

4.1.3 Adaptive Resonance Theory (ART) Networks

Adaptive Resonance Theory (ART) models are neural networks that perform clustering and can allow the number of clusters to vary. The major difference between ART and other clustering methods is that ART allows the user to control the degree of similarity between members of the same cluster by means of a user-defined constant called the *vigilance parameter*.

We outline in this section the simpler version of the ART network called ART1. The architecture of an ART1 network shown in Figure 4.3 consists of two layers of neurons. The F_1 neurons and the F_2 (cluster) neurons together with a reset unit control the degree of similarity of patterns or sample elements placed on the same cluster unit.

The ART1 networks accept only binary inputs. For continuous input, the ART2 network was developed with a more complex F_1 layer to accommodate continuous input.

The input layer, F_1 receives and holds the input patterns, the second layer, F_2 responds with a pattern associated with the given input. If this returned pattern is sufficiently similar to the input pattern, then there is a match. However, if the difference is substantial, then the layers communicate until a match is discovered, otherwise a new cluster of patterns is formed around the new input vector.

The training algorithm is shown in Figure 4.4 (Mehrotra et al. [142] or Fausett [69]).

The following example of Mehrotra et al. [142] illustrates the computations.

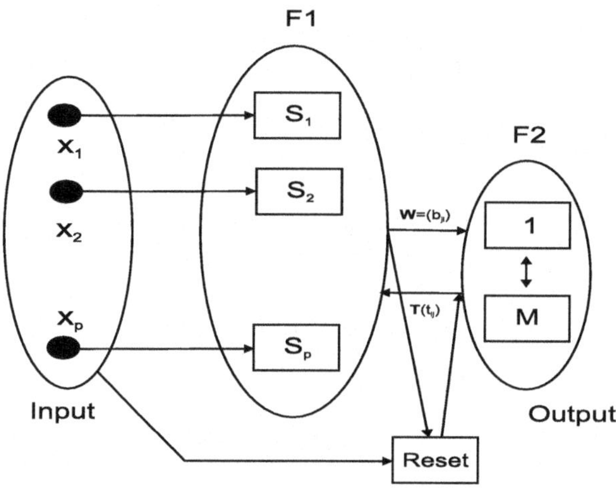

Figure 4.3
Architecture of ART1 network.

Consider a set of vectors $\{(1,1,0,0,0,0,1), (0,0,1,1,1,1,0), (1,0,1,1,1,1,0), (0,0,0,1,1,1,0)$, and $(1,1,0,1,1,1,0)\}$ to be clustered using the ART1 algorithm. Let the vigilance parameter be $\rho = 0.7$.

We begin with a single node whose top-down weights are all initialized to 1, i.e., $t_{l,1}(0) = 1$, and bottom-up weights are set to $b_{1,l}(0) = \frac{1}{8}$. Here $n = 7$ and initially $m = 1$. Given the first input vector, $(1,1,0,0,0,0,1)$, we compute

$$y_1 = \frac{1}{8} \times 1 + \frac{1}{8} \times 1 + \frac{1}{8} \times 0 + \ldots + \frac{1}{8} \times 0 + \frac{1}{8} \times 1 = \frac{3}{8}, \tag{4.9}$$

and y_1 is declared the uncontested winner. Since

$$\frac{\sum_{l=1}^{7} t_{l,1} x_l}{\sum_{l=1}^{7} x_l} = \frac{3}{3} = 1 > 0.7, \tag{4.10}$$

the vigilance condition is satisfied, and the updated weights are

$$b_{1,l}(1) = \begin{cases} \dfrac{1}{0.5+3} = \dfrac{1}{3.5} & \text{for } l = 1,2,7; \\ 0 & \text{otherwise.} \end{cases} \tag{4.11}$$

Likewise,

$$t_{l,1}(1) = t_{l,1}(0) x_l. \tag{4.12}$$

- Initialize each top-down weight $t_{l,j}(0) = 1$;
- Initialize each bottom-up weight $b_{j,l}(0) = \dfrac{1}{n+1}$;
- **While** the network has not stabilized, **do**
  1. Present a randomly chosen pattern $x = (x_1, \ldots, x_n)$ for learning.
  2. Let the active set A contain all nodes; calculate $y_j = b_{j,l}x_1 + \ldots + b_{j,n}x_n$ for each node $j \in A$;
     (a) Let j^* be a node in A with largest y_j, with ties being broken arbitrarily;
     (b) Compute $s^* = (s_1^*, \ldots, s_n^*)$ where $s_l^* = t_{l,j^*}x_l$;
     (c) Compare similarity between s^* and x with the given vigilance parameter ρ:
     if $\dfrac{\sum_{l=1}^{n} s_l^*}{\sum_{l=1}^{n} x_l} \leq \rho$ then remove j^* from set A
     else associate x with node j^* and update weights:
     $$b_{j^*,l}\,(\text{new}) = \frac{t_{l,j^*}\,(\text{old})\,x_l}{0.5 + \sum_{l=1}^{n} t_{l,j^*}\,(\text{old})\,x_l}$$
     $$t_{l,j^*}\,(\text{new}) = t_{l,j^*}\,(\text{old})\,x_l$$
     until A is empty or x has been associated with some node j;
  3. If A is empty, then create a new node whose weight vector coincides with the current input pattern x;

  end-while

Figure 4.4
Algorithm for updating weights in ART1.

These equations yield the following weight matrices.

$$B(1) = \left[\frac{1}{3.5}\ \frac{1}{3.5}\ 0\ 0\ 0\ 0\ \frac{1}{3.5}\right]^T, \qquad T(1) = \left[1\ 1\ 0\ 0\ 0\ 0\ 1\right]^T, \tag{4.13}$$

Now, we present the second sample $(0,0,1,1,1,1,0)$. This generates $y_1 = 0$, but the uncontested winner does not satisfy the vigilance threshold since $\sum_l t_{l,1}x_l / \sum_l x_l = 0 < 0.7$. A second node must hence be generated, with top-down weights identical to the sample, and bottom-up weights equal to 0 in the positions corresponding to the zeroes in the sample, and remaining new bottom-up weights are equal to $1/(0.5+0+0+1+1+1+1+0)$.

The new weight matrices are

$$B(2) = \begin{bmatrix} \frac{1}{3.5} & \frac{1}{3.5} & 0 & 0 & 0 & 0 & \frac{1}{3.5} \\ 0 & 0 & \frac{1}{4.5} & \frac{1}{4.5} & \frac{1}{4.5} & \frac{1}{4.5} & 0 \end{bmatrix}^T \tag{4.14}$$

and

$$T(2) = \begin{bmatrix} 1 & 1 & 0 & 0 & 0 & 0 & 1 \\ \\ 0 & 0 & 1 & 1 & 1 & 1 & 0 \end{bmatrix}^T.$$

When the third vector $(1, 0, 1, 1, 1, 1, 0)$ is presented to this network,

$$y_1 = \frac{1}{3.5} \text{ and } y_2 = \frac{4}{4.5} \tag{4.15}$$

are the node outputs, and the second node is the obvious winner. The vigilance test succeeds, because $\sum_{l=1}^{7} t_{l,2}x_l / \sum_{l=1}^{7} x_l = \frac{4}{5} \geq 0.7$. The second node's weights are hence adapted, with each top-down weight being the product of the old top-down weight and the corresponding element of the sample $(1, 0, 1, 1, 1, 1, 0)$, while each bottom-up weight is obtained on dividing this quantity by $0.5 + \sum_l t_{l,2}x_l = 4.5$. This results in no change to the weight matrices, so that $B(3) = B(2)$ and $T(3) = T(2)$.

When the fourth vector $(0, 0, 0, 1, 1, 1, 0)$ is presented to this network,

$$y_1 = \frac{1}{3.5} \text{ and } y_2 = \frac{3}{4.5} \tag{4.16}$$

are the node outputs, and the second node is the obvious winner. The vigilance test succeeds, because $\sum_{l=1}^{7} t_{l,2}x_l / \sum_{l=1}^{7} x_l = \frac{3}{3} \geq 0.7$. The second node's weights are hence adapted, with each top-down weight being the product of the old top-down weight and the corresponding element of the sample $(0, 0, 0, 1, 1, 1, 0)$, while each bottom-up weight is obtained on dividing this quantity by $0.5 + \sum_l t_{l,2}x_l = 3.5$. The resulting weight matrices are

$$B(4) = \begin{bmatrix} \frac{1}{3.5} & \frac{1}{3.5} & 0 & 0 & 0 & 0 & \frac{1}{3.5} \\ \\ 0 & 0 & 0 & \frac{1}{3.5} & \frac{1}{3.5} & \frac{1}{3.5} & 0 \end{bmatrix}^T, T(4) = \begin{bmatrix} 1 & 1 & 0 & 0 & 0 & 0 & 1 \\ \\ 0 & 0 & 0 & 1 & 1 & 1 & 0 \end{bmatrix}^T. \tag{4.17}$$

When the fifth vector $(1, 1, 0, 1, 1, 1, 0)$ is presented to this network,

$$y_1 = \frac{2}{3.5} \text{ and } y_2 = \frac{3}{4.5} \tag{4.18}$$

are the node outputs, and the second node is the obvious winner. The vigilance test fails, because $\sum_{l=1}^{7} t_{l,2}x_l / \sum_{l=1}^{7} x_l = \frac{3}{5} < 0.7$. The active set A is hence reduced to contain only the first node, which is the new winner (uncontested). The vigilance test fails with this node as well, with $\sum_l t_{l,1}x_l / \sum_{l=1}^{7} x_l = \frac{2}{5} \leq 0.7$. A third node is hence created, and the resulting weight matrices are

$$B(5) = \begin{bmatrix} \frac{1}{3.5} & \frac{1}{3.5} & 0 & 0 & 0 & 0 & \frac{1}{3.5} \\ 0 & 0 & 0 & \frac{1}{3.5} & \frac{1}{3.5} & \frac{1}{3.5} & 0 \\ \frac{1}{5.5} & \frac{1}{5.5} & 0 & \frac{1}{5.5} & \frac{1}{5.5} & 0 & \end{bmatrix}^T, \quad T(5) = \begin{bmatrix} 1 & 1 & 0 & 0 & 0 & 0 & 1 \\ 0 & 0 & 0 & 1 & 1 & 1 & 0 \\ 1 & 1 & 0 & 1 & 1 & 1 & 0 \end{bmatrix}^T. \tag{4.19}$$

We now cycle through all the samples again. This can be performed in random order, but we opt for the same sequence as in the earlier cycle. After the third vector is present, again, the weight matrices are modified to the following:

$$B(8) = \begin{bmatrix} \frac{1}{3.5} & \frac{1}{3.5} & 0 & 0 & 0 & 0 & \frac{1}{3.5} \\ 0 & 0 & 0 & \frac{1}{3.5} & \frac{1}{3.5} & \frac{1}{3.5} & 0 \\ \frac{1}{4.5} & 0 & 0 & \frac{1}{4.5} & \frac{1}{4.5} & \frac{1}{4.5} & 0 \end{bmatrix}^T, \quad T(8) = \begin{bmatrix} 1 & 1 & 0 & 0 & 0 & 0 & 1 \\ 0 & 0 & 0 & 1 & 1 & 1 & 0 \\ 1 & 0 & 0 & 1 & 1 & 1 & 0 \end{bmatrix}^T. \tag{4.20}$$

Subsequent presentations of the samples do not result in further changes to the weights, and $T(8)$ represents the prototypes for the given samples. The network has thus stabilized. For further details see Mehrotra et al. [142].

In what follows, we describe some applications of ART neural networks.

Santos [184] and Santos et al. [186] used neural networks and classification trees in the diagnosis of smear negative pulmonary tuberculosis (SPNT), which accounted for 30% of the reported cases of pulmonary tuberculosis. The data was obtained from 136 patients from the health care unit of the Universidade Federal do Rio de Janeiro teaching hospital referred from March 2001 through September 2002.

Only symptoms and physical condition were used for constructing the neural networks and classification tree. The covariate vector contained 3continuous variables and 23 binary variables.

In this application, an ART neural network identified three groups of patients. In each group the diagnostic was obtained from one hidden layer feedforward network. The neural networks showed sensitivity of 71% to 84% and specificity of 61% to 83% and performed slightly better than the classification tree. Statistical models in literature based on laboratory results showed sensitivity of 49% to 100% and specificity of 16 to 86%. The neural networks of Santos [184] and Santos et al. [186] used only clinical information.

Rozenthal [182] and Rozenthal et al. [183] applied an ART neural network to analyze data from 53 schizophrenic (not addicted, physically capable, below age 50) patients who met the Diagnostic and Statistical Manual for Mental Disorders (DSM-IV) criteria and submitted to neuropsychological tests. Schizophrenia patients exhibit at least two functional symptom patterns: those who have hallucinations, disorderly thoughts and low self-esteem (negative dimension); those who have poor speech and disorderly thoughts (disorganized dimension). This application of the neural network (along with classical statistical clustering for comparison) indicated two important clusters. The low IQ and negative dimension cluster remained stable when the number of clusters was increased. This cluster seemed to act as an attractor that impacted the more severe cases and did not respond readily to treatment. Tables 4.1 though 4.4 present the results of the study.

Table 4.1
Presentation of the sample according to level of instruction, IQ and age of onset.

Level of Instruction	No. Patients ($n = 53$)	IQ ($m \pm dp$)	Age of Onset ($m \pm dp$)
College	22	84.91 ± 07.59	21.64 ± 5.58
High school	19	80.84 ± 0.07	19.5 ± 6.13
Elementary	12	75.75 ± 08.88	16.67 ± 3.89

Table 4.2
Cluster patterns according to neuropsychological findings.

	P1	P2	P3	P4	P5	P6	P7	P8	P9	P10	P11	RAVLT[1]
Group I ($n = 20$)	77.6	7.7	7.0	-0.05	4.2	11.2	2.7	0.05	4.2	-1.8	-4.2	8.3
Group II ($n = 3$)	84	2.0	24.1	-3.1	4.6	11.6	2.0	100	5.5	-1.8	9.1	11.5

[1] Rey Auditory Verbal Learning Test (RAVLT).

Table 4.3
Three cluster patterns according to neuropsychological findings.

	P1	P2	P3	P4	P5	P6	P7	P8	P9	P10	P11	RAVLT[1]
Group I ($n = 20$)	77.6	7.7	7.0	-0.05	4.2	11.2	2.7	0.05	4.2	-1.8	-4.2	8.3
Group IIa ($n = 20$)	84.7	-0.5	57.1	-14.1	4.5	10.9	1.9	1	5.0	-1.7	-9.1	11.4
Group IIb ($n = 13$)	84.9	3.7	29	-4.0	4.7	12.1	2.1	1	5.8	-1.8	20.8	11.6

[1] Rey Auditory Verbal Learning Test (RAVLT).

Chiu et al. [39] developed a self-organizing cluster system for the arterial pulse based on the ART neural network. The technique provides at least three novel diagnostic tools in the clinical neurophysiology laboratory. First, the pieces affected by unexpected artificial motions (i.e., physical disturbances) can be determined easily

Table 4.4
Four cluster patterns according to neuropsychological findings.

	P1	P2	P3	P4	P5	P6	P7	P8	P9	P10	P11	RAVLT[1]
Group I ($n = 20$)	77.6	7.7	7.0	-0.05	4.2	11.2	2.7	0.05	4.2	-1.8	-4.2	8.3
Group IIc ($n = 14$)	80.86	0.81	12.2	6.8	4.7	10.9	2.2	1	6.0	-2.0	0.9	11.4
Group IId ($n = 11$)	85.1	-1.4	51.90	-12.0	4.9	11.6	2.0	1	4.7	-1.7	-21.2	11.2
Group IIe ($n = 8$)	90.0	7.7	43.0	-6.7	4.2	12.5	1.7	1	5.6	-1.6	55.4	12

[1] Rey Auditory Verbal Learning Test (RAVLT).

by the ART2 neural network according to a status distribution plot. Second, a few categories will be created after applying the ART2 network to the input patterns (i.e., minimal cardiac cycles). These pulse signal categories can be useful to physicians for diagnosis in conventional clinical uses. Third, the status distribution plot provides an alternative method to assist physicians in evaluating the signs of abnormal and normal automatic control. The method has shown clinical applicability for the examination of the autonomic nervous system.

Ishihara et al. [98] built a Kansei engineering system based on an ART neural network to assist designers in creating products fit for consumers' underlying needs, with minimal effort. Kansei engineering is a technology for translating human feeling into product design. Statistical models (linear regression, factor analysis, multidimensional scaling, centroid clustering, etc.) are used to analyze the feeling-design relationship and build rules for Kansei engineering systems. Although reliable, they are time and resource consuming and require expertise in relation to mathematical constraint. They found that their ART neural network enables quick, automatic rule building in Kansei Engineering systems. The categorization and feature selection of their rule were also slightly better than results with conventional multivariate analysis.

4.1.4 Self-Organizing Maps (SOM) Networks

Self-organizing maps, sometimes called topologically ordered maps or Kohonen self-organizing feature maps are closely related to multidimensional scaling methods. The goal is to represent all points in the original space by points in a smaller space, such that distance and proximity relations are preserved as much as possible.

There are m cluster units arranged in one- or two-dimensional array.

The weight vector for a cluster neuron serves as an example of the input patterns associated with that cluster. During the self-organizing process, the cluster neuron whose weight vector matches the input pattern most closely (usually, the square of the minimum Euclidean distance) is chosen as the winner. The winning neuron and its neighboring units (in terms of the topology of the cluster neurons) update their weights in the usual topology in two-dimensions as shown in Figure 4.5).

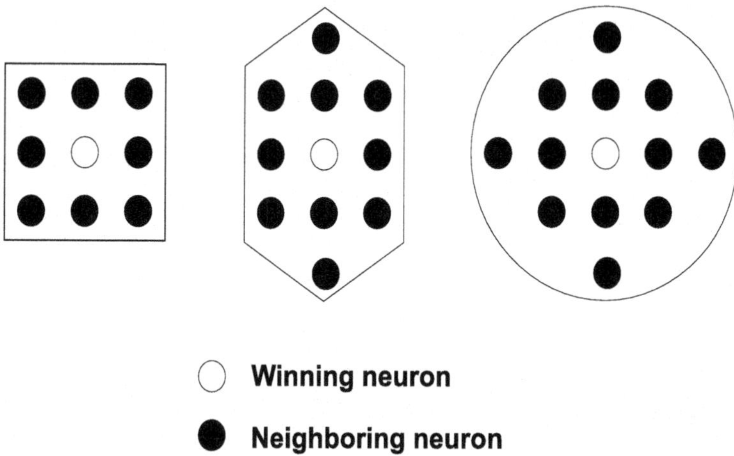

Figure 4.5
Topology of neighboring regions.

The architecture of the Kohonen SOM neural network for one and two-dimensional arrays as well as their topologies are shown in Figures 4.6 through 4.10. Figure 4.11 presents the weight updating algorithm for the Kohonen SOM neural networks (from [69]) and Figure 4.12, the Mexican Hat interconnection for neurons in the algorithm.

Braga et al. [27] utilized 1000 samples with equal probability from two bivariate normal distributions with variances 1 and mean vectors $\mu_1 = (4, 4)$,$\mu_2 = (12, 12)$. The distribution and the data are shown in Figures 4.13 and 4.14.

The topology was such that each neuron had 4 other neighboring neurons. The resulting map was expected to preserve the statistical characteristics, i.e., the visual inspection of the map should indicate how the data is distributed.

Figure 4.15 gives the initial conditions (weights). Figures 4.16, 4.17, and 4.18 present the map after some interactions.

Lourenço [130] applied a combination of SOM networks, fuzzy set and classical forecasting methods for the short-term electricity load prediction. The method establishes the various load profiles and processes climatic variables in a linguistic way, as well as those from the statistical models. The final model includes a classifier scheme, a predictive scheme, and a procedure to improve the estimations. The classifier is implemented via a SOM neural network. The forecaster uses statistical forecasting techniques (moving average, exponential smoothing, and ARMA type

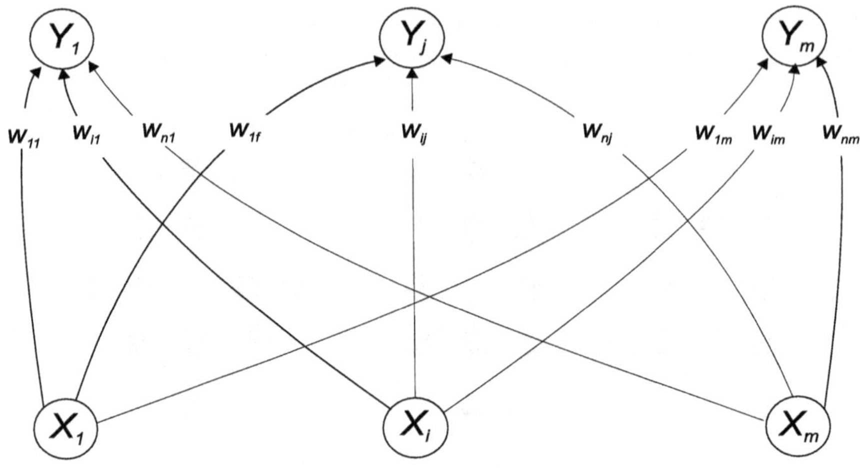

Figure 4.6
Kohonen self-organizing map.

* * * {* (* [#] *) *} * *

{ } R=2 ( ) R=1 [ ] R=0

Figure 4.7
Linear array of cluster units.

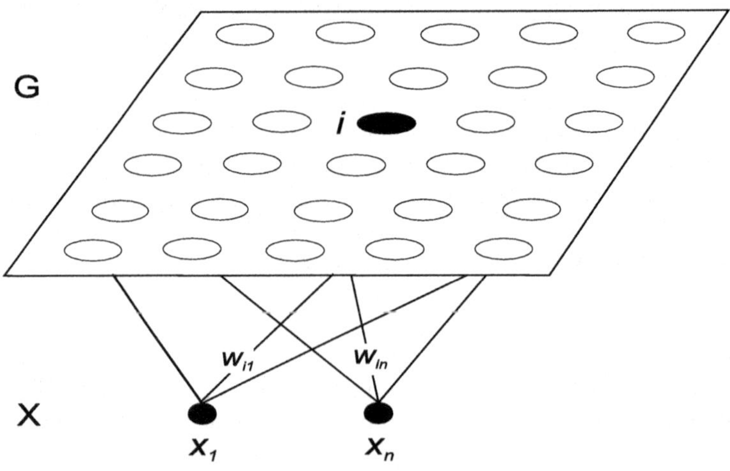

Figure 4.8
The self-organizing map architecture.

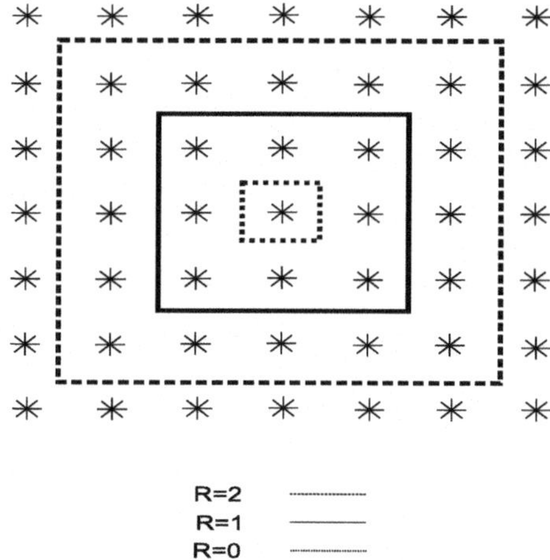

Figure 4.9
Neighborhoods for rectangular grid.

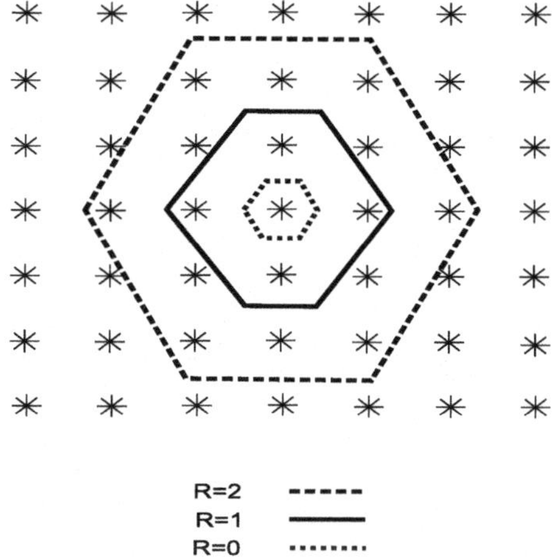

Figure 4.10
Neighborhoods for hexagonal grid.

Step 0 Initialize weights w_{ij}
Set topological neighborhood parameters.
Set learning rate parameters.
Step 1 While stopping condition is false, do Steps 2-8.
Step 2 For each input vector $\boldsymbol{x}$, do Steps 3-5.
Step 3 For each j, compute:
$D(j) = \sum_i (w_{ij} - x_i)^2$
Step 4 Find index J such that $D(J)$ is minimum.
Step 5 For all units within a specified neighborhood of J and for all i:
$w_{ij}(\text{new}) = w_{ij}(\text{old}) + \alpha[x_i - w_{ij}(\text{old})]$
Step 6 Update learning rate.
Step 7 Reduces radius of topological neighborhood at specified times.
Step 8 Test stopping condition.

Figure 4.11
Algorithm for updating weights in SOM.

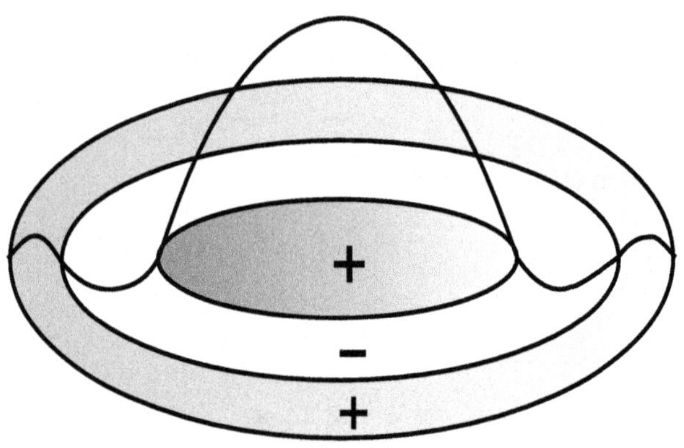

Figure 4.12
Mexican Hat interconnections.

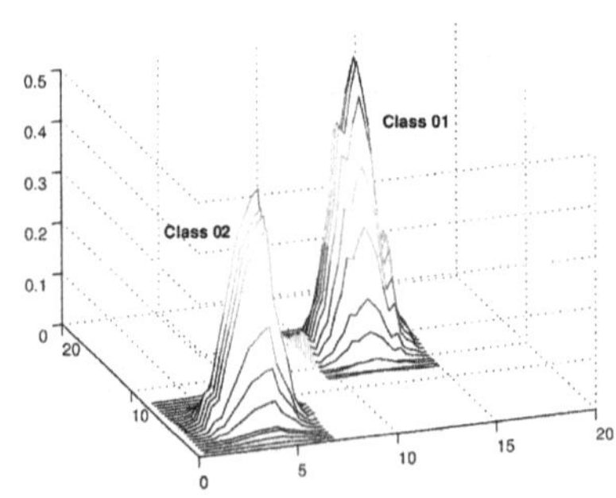

Figure 4.13
Normal density function of two clusters.

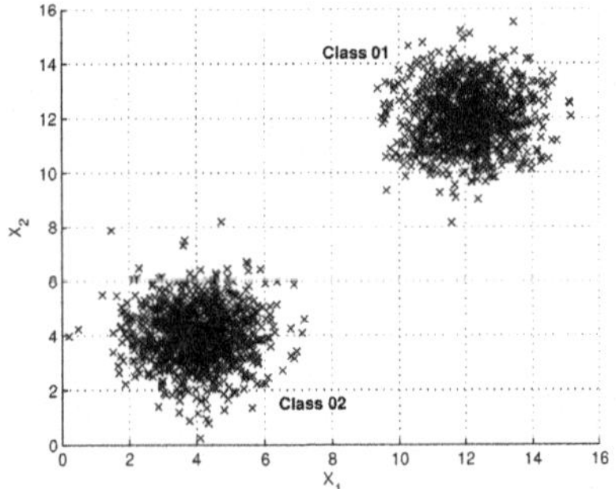

Figure 4.14
Sample from normal cluster for SOM training.

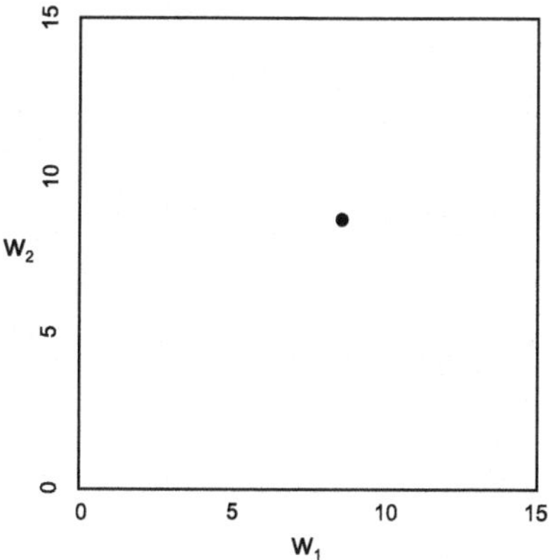

Figure 4.15
Initial weights.

Figure 4.16
SOM weights with 50 iterations.

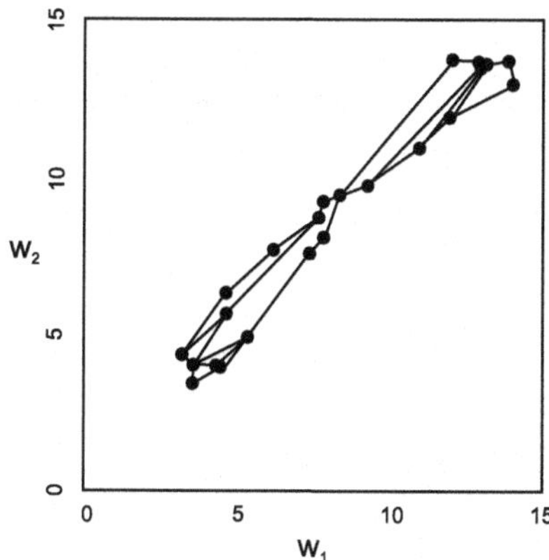

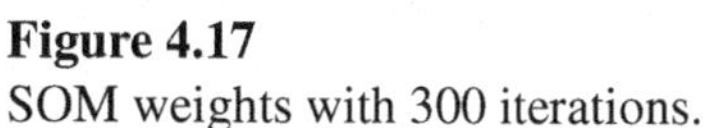

Figure 4.17
SOM weights with 300 iterations.

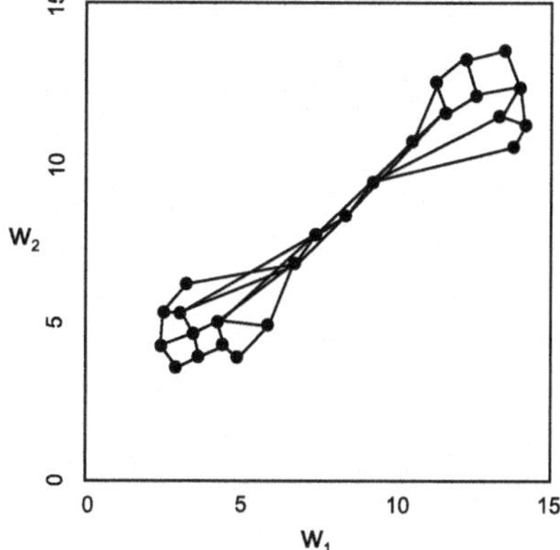

Figure 4.18
SOM weights with 4500 iterations.

models). A fuzzy logic procedure uses climatic variables to improve the forecast. (See also Kartalopulos [108].)

The complete system is shown in Figure 4.19.

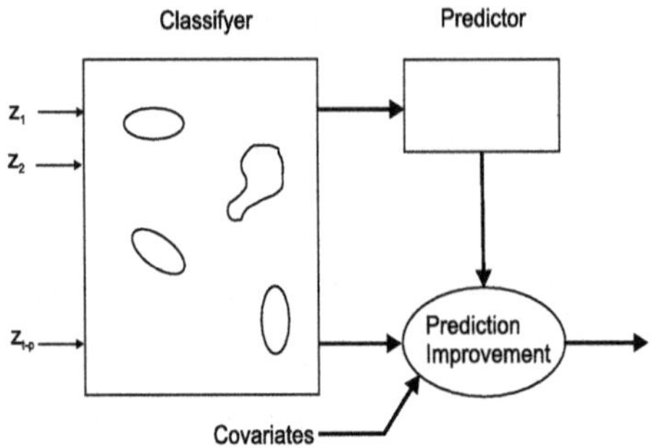

Figure 4.19
Classifier forecast system.

The classifier used a SOM network with 12 neurons displaced in four rows and three columns. The classification of patterns is shown in Tables 4.5 and 4.6 for the year of 1993.

Table 4.5
Week days associated with the SOM neurons.

^{1}Sunday	^{2}Sunday	^{3}Saturday
^{4}Saturday	^{5}Saturday	^{6}Monday to Friday
^{7}Monday to Friday	^{8}Monday to Friday	^{9}Monday to Friday
^{10}Monday to Friday	^{11}Monday to Friday	^{12}Monday to Friday

Table 4.6
Seasons associated with the SOM neurons.

^{1}	^{2}	^{3}
^{4}	^{5}	^{6}Winter
^{7}Summer	^{8}Spring	^{9}Winter
^{10}Summer	^{11}Autumn	^{12}Winter

The proposed method showed improvement over the method currently used by the Brazilian Electric System to predict loads for peak hours.

Draghici and Potter [61] predicted HIV drug resistance. Since drug resistance is an important factor influencing the failure of HIV therapies, the ability to predict the resistance of HIV protease mutants may lead to development of more effective and longer-lasting treatment regimens.

They predicted the HIV resistance to two current protease inhibitors. The problem was approached from two perspectives. First, a predictor was constructed based on the structural features of the HIV protease-drug inhibitor complex. A particular structure was represented by its list of contacts between the inhibitor and the protease. Next, a classifier was constructed based on the sequence data of various drug resistant mutants. In both cases, SOM networks were first used to extract the important features and cluster the patterns in an unsupervised manner. This was followed by subsequent labelling based on the known patterns in the training set.

The prediction performance of the classifiers was measured by cross validation. The classifier using the structure information correctly classified previously unseen mutants with an accuracy between 60% and 70%. Several architectures were tested on the more abundant sequence of 68% and a coverage of 69%. Multiple networks were then combined into various majority voting schemes. The best combination yielded an average of 85% coverage and 78% accuracy on previously unseen data. This is more than two times better than the 33% accuracy expected from a random classifier.

Hsu et al. [90] describe an unsupervised dynamic hierarchical self-organizing approach, which suggests an appropriate number of clusters, to perform class discovery and marker gene identification in microarray data. In the process of class discovery, the proposed algorithm identifies corresponding sets of predictor genes that best distinguish one class from other classes. The approach integrates merits of hierarchical clustering with robustness against noise generated from self-organizing approaches.

The proposed algorithm applied to DNA microarray data sets of two types of cancers has demonstrated its ability to produce the most suitable number of clusters. Further, the corresponding marker genes identified through the unsupervised algorithm also have a strong biological relationship to the specific cancer class. The algorithm tested on leukemia microarray data, which contains three leukemia types, was able to determine three major clusters and one minor cluster. Prediction models built for the four clusters indicate that the prediction strength for the smaller cluster is generally low, therefore labelled as an uncertain cluster. Further analysis shows that the uncertain cluster can be further subdivided, and the subdivisions are related to two of their original clusters. Another test performed using colon cancer microarray data has automatically derived two clusters, which is consistent with the number of classes in data (cancerous and normal).

Mangiameli et al. [138] provide a comparison of the performance of SOM network and seven hierarchical cluster methods: single linkage, complete linkage, average linkage, centroid method, Ward's method, two stage density and k-nearest neighbor. A total of 252 empirical data sets was constructed to simulate several levels of imperfections that include dispersion, aberrant observations, irrelevant variables, non-uniform clusters. More details are shown in Tables 4.7 through 4.9.

The experimental results are unambiguous; the SOM network is superior to all seven hierarchical clustering algorithms commonly used today. Furthermore, the per-

Table 4.7
Data set design.

Data set	Design factors	No. of data sets
Base data	2,3,4 or 5 clusters; 4,6 or 8 variables; low, med or high dispersion	36
Irrelevant variables	Basic data plus 1 or 2 irrelevant variables	72
Cluster density	Basic data plus 10% or 60% density	72
Outliers	Basic data plus 10% or 20% outliers	72

Table 4.8
Average distance between clusters.

No. of clusters	Average distance (units)
2	5.03
3	5.31
4	5.33
5	5.78

Table 4.9
Basis for data set construction.

Level of dispersion	Average cluster standard deviation (units)	Standard deviation range
High	7.72	3
Medium	3.72	6
Low	1.91	12

formance of the SOM network is shown to be robust across all these data imperfections. The SOM superiority is maintained across a wide range of "messy data" conditions that are typical of empirical data sets. Additionally, as the level of dispersion in the data increases, the performance advantage of the SOM network relative to the hierarchical clustering methods increases to a dominant level.

The SOM network is the most accurate method for 191 of the 252 data sets tested, which represents 75.8% of the data. The SOM network ranks first or second in accuracy in 87.3% of the data sets. For the high dispersion data sets, the SOM network is most accurate 90.2% of the time and is ranked first or second 91.5% of the time. The SOM network frequently has average accuracy levels of 85% or greater, while other techniques average between 35% and 70%. Of the 252 data sets investigated, only six data sets resulted in poor SOM results. These six data sets occurred at low levels of data dispersion, with a dominant cluster containing 60% or more of the observations and four or more total clusters. Despite the relatively poor performance at these data conditions, the SOM correctly classified an average of 72% of observations.

These findings seem to contradict the poor performance of SOM networks compared to k-means clustering method applied in 108 data sets reported by Balakrishnan et al. [13].

Pereira et al. [164] applied the SOM network to gamma ray burst and suggested the existence of 5 clusters of bursts. This result was confirmed by a feedforward network and principal component analysis.

4.2 Dimensional Reduction Networks

In this section, we will present some neural networks designed to deal with the problem of dimensional reduction of data, see Tura [212].

In several occasions, it is useful and even necessary to first reduce the dimension of the data to a manageable size, keeping as much of the original information as possible, and then to proceed with the analysis.

Sometimes, a phenomenon that is in appearance high-dimension is actually governed by few variables (sometimes called "latent variables" or "factors"). The redundancy, presented in the data collected related to the phenomenon, may arise, for example, because:

- Many of the variables collected may be irrelevant.
- Many of the variables will be correlated (therefore some redundant information is contained in the data), and a new set of uncorrelated or independent variables should be found.

An important reason to reduce the dimension of the data is that some authors call: "the curse of dimensionality and the empty space phenomenon". The curse of dimensionality phenomenon refers to the fact that in the absence of simplifying assumptions, the sample size needed to make inferences with a given degree of accuracy grows exponentially with the number of variables. The empty space phenomenon responsible for the curse of dimensionality is that high-dimensional spaces are inherently sparse. For example, for a one dimensional standard normal $N(0, 1)$, about 70% of the mass is at points contained in the interval (sphere of radius of one standard deviation around the mean zero). For a 10-dimensional $N(0, I)$, the same (hyper) sphere contains only $0, 02\%$ of the mass, and a radius of more than 3 standard deviations is needed to contain 70%.

A related concept in dimensional reduction methods is the intrinsic dimension of a sample, which is defined as the number of independent variables that explain satisfactorily the phenomenon. This is a loosely defined concept, and a trial and error process is usually used to obtain a satisfactory value for it.

There are several techniques to dimensional reduction, and they can be classified in two types: with and without exclusion of variables. The main techniques are:

- Selection of variables:
  - Expert opinion;
  - Automatic methods: all possible regressions, best subset regression, backward elimination, stepwise regression, etc.
- Using the original variables:
  - Principal components;
  - Factor analysis;
  - Correspondence analysis;
  - Multidimensional scaling;
  - Nonlinear principal components;
  - Independent component analysis;
  - Others.

We are concerned in this book with the latter group of techniques (some of which have already been described: projection pursuit, GAM, multidimensional scaling, etc.). For a general view, see also Correia-Perpinan [50].

Now, we turn to some general features and common operations on the data matrix to be used in these dimensional reduction methods.

4.2.1 Basic Structure of Data Matrix

A common operation in most dimensional reduction techniques is the decomposition of a matrix of data into its basic structure (see Clausen [46]).

We refer to the data set, the data matrix as $\boldsymbol{X}$, and the specific observation of variable j in subset i, as x_{ij}. The dimension of $\boldsymbol{X}$ is $(n\ timesp)$, corresponding to n observations of p variables $(n > p)$.

Any data matrix can be decomposed into its characteristic components:

$$\boldsymbol{X}_{n\,\mathrm{x}\,p} = \boldsymbol{U}_{n\,\mathrm{x}\,p}\boldsymbol{d}_{p\,\mathrm{x}\,p}\boldsymbol{V}^{'}_{p\,\mathrm{x}\,p}, \tag{4.21}$$

which is called "single value decomposition", i.e., SVD.

The $\boldsymbol{U}$ matrix "summarizes" the information in the rows of $\boldsymbol{X}$, the $\boldsymbol{V}$ matrix "summarizes" information in the columns of $\boldsymbol{X}$, the $\boldsymbol{d}$ matrix is a diagonal matrix, whose diagonal entries are the singular values and are weights indicating the relative importance of each dimension in $\boldsymbol{U}$ and $\boldsymbol{V}$, and are ordered from largest to smallest.

If $\boldsymbol{X}$ does not contain redundant information the dimension of $\boldsymbol{U}$, $\boldsymbol{V}$ and $\boldsymbol{d}$ will be equal to the minimum dimension p of $\boldsymbol{X}$. The dimension is called rank.

If $\boldsymbol{S}$ is a symmetric matrix the basic structure is:

$$\boldsymbol{S} = \boldsymbol{U}\boldsymbol{d}\boldsymbol{V}^{'} = \boldsymbol{U}\boldsymbol{d}\boldsymbol{U}^{'} = \boldsymbol{V}\boldsymbol{d}\boldsymbol{V}^{'}. \tag{4.22}$$

Decomposition of $\boldsymbol{X}$, $\boldsymbol{X}.\boldsymbol{X}^{'}$ or $\boldsymbol{X}^{'}.\boldsymbol{X}$ reveals the same structure:

$$\boldsymbol{X}_{nxp} = \boldsymbol{U}\boldsymbol{d}\boldsymbol{V}^{'} \tag{4.23}$$

$$\boldsymbol{X}\boldsymbol{X}^{'}_{(nxn)} = \boldsymbol{U}\boldsymbol{d}^{2}\boldsymbol{U}^{'} \tag{4.24}$$

$$\boldsymbol{X}\boldsymbol{X}^{'}_{(pxp)} = \boldsymbol{V}\boldsymbol{d}^{2}\boldsymbol{V}^{'}. \tag{4.25}$$

A rectangular matrix $\boldsymbol{X}$ can be made into symmetric matrices ($\boldsymbol{X}\boldsymbol{X}^{'}$ and $\boldsymbol{X}^{'}\boldsymbol{X}$) and their eigenstructure can be used to obtain the structure of $\boldsymbol{X}$. If $\boldsymbol{X}\boldsymbol{X}^{'} = \boldsymbol{U}\boldsymbol{D}$ and $\boldsymbol{X}^{'}\boldsymbol{X} = \boldsymbol{V}\boldsymbol{D}\boldsymbol{V}^{'}$ then $\boldsymbol{X} = \boldsymbol{U}\boldsymbol{D}^{\frac{1}{2}}\boldsymbol{V}$.

The columns of $\boldsymbol{U}$ and $\boldsymbol{V}$ corresponding to the eigenvalues D_i are also related. If U_i is an eigenvector of $\boldsymbol{X}\boldsymbol{X}^{'}$ corresponding to D_i, then the corresponding eigenvector of $\boldsymbol{X}^{'}\boldsymbol{X}$ corresponding to D_i is equal or proportional to $\boldsymbol{U}^{'}_{i}\boldsymbol{X}$.

Some dimensional reduction techniques share the same decomposition algorithm SVD. They differ only in the predecomposition transformation of the data and their postdecomposition transformation of the latent variables. The following transformations will be used:

- Deviation from the mean

$$x_{ij} = x_{ij} - \bar{X}_j, \tag{4.26}$$

where x_j is the column (variable j) mean.

- Standardization

$$z_{ij} = \frac{x_{ij} - \bar{X}_j}{S_j}, \tag{4.27}$$

where $\bar{X}_j$ and S_j are the column (variable j) mean and standard deviation.

- Unit length

$$x^{*}_{ij} = \frac{x_{ij}}{(\sum x^{2}_{ij})^{\frac{1}{2}}}. \tag{4.28}$$

- Double-centering

$$x^*_{ij} = x_{ij} - \bar{X}_i - \bar{X}_j + \bar{X}_{ij} \tag{4.29}$$

obtained by subtracting the row and column means and adding back the overall mean.

- For categorical and frequency data

$$x^*_{ij} = \frac{x_{ij}}{\left(\sum_i x_{ij} \sum_j x_{ij}\right)^{\frac{1}{2}}} = \frac{x_{ij}}{(x_{\bullet j} x_{i \bullet})^{\frac{1}{2}}} \tag{4.30}$$

From these transformations the following matrices are obtained:

- Covariance matrix

$$C = \frac{1}{n}(\boldsymbol{X} - \bar{\boldsymbol{X}})'(\boldsymbol{X} - \bar{\boldsymbol{X}}') \tag{4.31}$$

from the mean corrected data.

- R-correlation matrix

$$R = \frac{1}{n}\boldsymbol{Z}_c'\boldsymbol{Z}_c, \tag{4.32}$$

where $\boldsymbol{Z_c}$ indicates the matrix $\boldsymbol{X}$ standardized within columns in the number of observations.

- Q-correlation matrix

$$Q = \frac{1}{p}\boldsymbol{Z}_r'\boldsymbol{Z}_r, \tag{4.33}$$

where $\boldsymbol{Z_r}$ indicates the matrix $\boldsymbol{X}$ standardized within rows, and p is the number of variables.

4.2.2 Mechanics of Some Dimensional Reduction Techniques

Now, we outline the relation of the SVD algorithm to the dimensional reduction algorithm, some of which can be computed using neural networks.

4.2.2.1 Principal Components Analysis (PCA)

When applying PCA in a set of data the objectives are:

- To obtain from the original variables a new set of variables (factors, latent variables), which are uncorrelated.

- To hope that a few of these new variables will account for most of the variation of the data; that is the data can be reasonably represented in a lower dimension and keeping most of the original information.
- These new variables can be reasonable interpreted.

The procedure is performed by applying an SVD on the C-correlation matrix or in the R-correlation matrix. Since PCA is not invariant to transformation, usually the R matrix is used (see also Dunteman [62] and Jackson [100]).

4.2.2.2 Nonlinear Principal Components

Given the data matrix $\boldsymbol{X}$, this procedure consists in performing the SVD algorithm in some function $\phi(\boldsymbol{X})$.

In the statistical literature, an early reference is Gnandesikan [79] and it is closely related to kernel principal components, principal curves, and the informax method in independent component analysis. The latter will be outlined later.

4.2.2.3 Factor Analysis (FA)

Factor analysis has also the aim of reducing the dimensionality of a variable set and the aim of representing a set of variables in terms of a smaller number of hypothetical variables (see [112]).

The main differences between PCA and FA are:

- PCA decomposes the total variance. In the case of standardized variables, it produces a decomposition of $\boldsymbol{R}$. In contrast, FA finds a decomposition of the reduced matrix $\boldsymbol{R}-\boldsymbol{U}$, where $\boldsymbol{U}$ is a diagonal matrix of the "unique" variances associated with the variables. A unique variance is the part of each variable's variance that has nothing in common with remaining $p-1$ variables.
- PCA is a procedure to decompose the correlation matrix without regard to an underlying model. In contrast, FA has an underlying model that rests on a number of assumptions, including normality as the distribution for the variables.
- The emphasis in FA is in explaining X_i as a linear function of hypothetical unobserved common factors plus a factor unique to that variable, while the emphasis in PCA is expressing the principal components as a linear function of the X_i's. Contrary to PCA, the FA model does not provide a unique transformation from variables to factors.

4.2.2.4 Correspondence Analysis (CA)

Correspondence analysis represents the rows and columns of a data matrix as points in a space of low dimension, and it is particularly suited to (two-way) contingency tables (see Clausen [46] and Weller and Rommey [217]).

Log-linear model (LLM) is also a method widely used for analyzing contingency tables. The choice of method, CA or LLM depends of the type of data to be analyzed and on what relations or effects are of most interest. A practical rule for deciding when each method is to be preferred is: CA is very suitable for discovering the inherent structures of contingency tables with variables containing a large number of categories. LLM is particularly suitable for analyzing multivariate contingency tables, where the variables contain few categories. LLM analyses mainly the interrelationships between a set of variables, whereas correspondence analysis examines the relations between the categories of the variables.

In CA, the data consists of an array of frequencies with entries f_{ij}, or as a matrix F.

- The first step is to consider the transformation

$$h_{ij} = \frac{f_{ij}}{\sqrt{f_{i\bullet} f_{\bullet j}}}, \tag{4.34}$$

  where h_{ij} is the entry for a given cell, f_{ij} the original cell frequency, $f_{i\bullet}$ the total for row i and $f_{\bullet j}$ the total for column j.

- The second step is to find the basic structure of the normalized matrix $\boldsymbol{H}$ with elements (h_{ij}) using SVD. This procedure summarizes row and column vectors, $\boldsymbol{U}$ and $\boldsymbol{V}$ of $\boldsymbol{H}$ and a diagonal matrix of singular values, **d**, of $\boldsymbol{H}$. The first singular value is always one and the successive values constitute singular values or canonical correlations.

- The third and final step is to rescale the row ($\boldsymbol{U}$) and column ($\boldsymbol{V}$) vectors to obtain the canonical scores. The row ($\boldsymbol{X}$) and column ($\boldsymbol{Y}$) canonical scores are obtained as

$$X_i = \boldsymbol{U}_i \sqrt{\frac{f_{\bullet\bullet}}{f_{i\bullet}}}, \tag{4.35}$$

  and

$$Y_i = \boldsymbol{V}_i \sqrt{\frac{f_{\bullet\bullet}}{f_{\bullet j}}}. \tag{4.36}$$

4.2.2.5 Multidimensional Scaling

Here, we will describe the classical or metric method of multidimensional scaling, which is closely related to the previous dimensional reduction method, since it is also based on SVD algorithm. Alternative mapping and ordinal methods will not be outlined.

In multidimensional scaling, we are given the distances d_{rs} between every pair of observations. Given the data matrix $\boldsymbol{X}$, there are several ways to measure the distance between pairs of observations. Since many of them produce measures of distances, which do not satisfy the triangle inequality, the term dissimilarity is used.

For continuous data $\boldsymbol{X}$, the most common choice is the Euclidean distance $d^2 = \boldsymbol{X}\boldsymbol{X}'$. This depends on the scale in which the variables are measured. One way out is to use Mahalanobis distance with respect to the covariance matrix $\hat{\Sigma}$ of the observations.

For categorical data, a commonly used dissimilarity is based on the simple matching coefficient, i.e., the proportion c_{rs} of features, which are common to the two observations r and s. As this is between zero and one, the dissimilarity is found to be $d_{rs} = 1 - c_{rs}$.

For ordinal data, we use their rank as if they were continuous data after rescaling to the range $(0, 1)$ so that every original feature is given equal weight.

For a mixture of continuous, categorical and ordinal features a widely used procedure is: for each feature define a dissimilarity d^f_{rs} and an indicator I^f_{rs}, which is one only if feature f is recorded for both observations. Further, $I^f_{rs} = 0$ if we have a categorical feature and an absence-absence. Then

$$d_{rs} = \frac{\sum_f I^f_{rs} d^f_{rs}}{\sum I^f_{rs}}. \tag{4.37}$$

In the classical or metric method of multidimensional scaling, also known as principal coordinate analysis, we assume that the dissimilarities were derived as Euclidean distances between n points in p dimensions. The steps are as follows:

- From the data matrix obtain the matrix of distances $\boldsymbol{T}$ ($\boldsymbol{X}\boldsymbol{X}'$ or $\boldsymbol{X}\Sigma\boldsymbol{X}'$).
- Double center the dissimilarity matrix $\boldsymbol{T}$, say $\boldsymbol{T}^*$.
- Decompose the matrix $\boldsymbol{T}^*$ using the SVD algorithm.
- From the p (at most) non-zero vectors represent the observations in the space of two (or three) dimensions.

For more details see Cox and Cox [52].

4.2.2.6 Independent Component Analysis (ICA)

The ICA model will be easier to explain, if the mechanics of a cocktail party are first described. At a cocktail party, there are speakers holding conversations while at the same time, microphones (also called underlying sources of components) record the speakers' conversations.

At a cocktail party, there are p microphones that record or observe m partygoers or speakers at n instants. This notation is consistent with traditional multivariate statistics. The observed conversation consists of mixtures of true unobserved conversations. The microphones do not record the conversations in isolation; the conversations are mixed. The problem is to unmix or recover the original conversation from the recorded mixed conversations.

The relation of ICA with other methods of dimension reduction is shown in Figure 4.20 as discussed in Hyvärinen [94]. The lines show close connections, and the text next to the lines shows the assumptions needed for the connection.

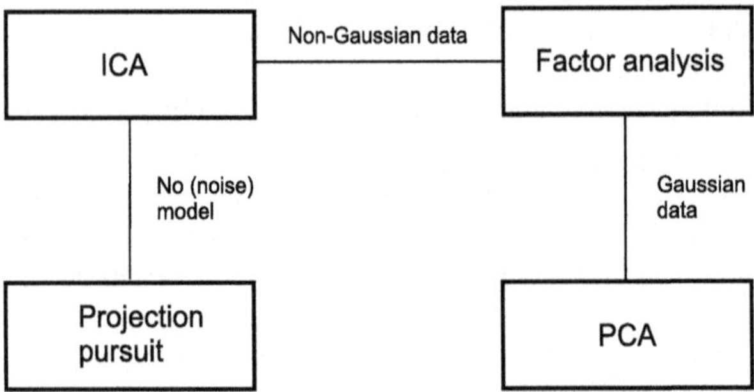

Figure 4.20
Relationship of ICA with other dimension reduction methods. The lines show close connections and the text next to the lines shows the assumptions needed for the connections..

Before we formalize the ICA method, we should mention that the roots of basic ICA can be traced back to the work of Darmois in the 1950s (see Darmois [54]) and Rao in the 1960s (see Kagan et al. [104]) concerning the characterization of random variables in linear structures. Only recently this has been recognized by ICA researchers (Comon [47], Jutten and Taleb [103], Fiori [71]), who regret that the delay may have impeded progress.

For the general problem of source separation, we use the statistical latent variables model. Suppose that we have p linear mixtures of m independent components given by

$$\begin{bmatrix} x_{i1} \\ \dots \\ x_{ip} \end{bmatrix} = \begin{bmatrix} a_{11}.s_{i1} + \ldots + a_{1m}.s_{im} \\ \dots \\ a_{p1}.s_{i1} + \ldots + a_{pm}.s_{im} \end{bmatrix} = \begin{bmatrix} a_{11} & \dots & a_{1m} \\ & & \\ a_{p1} & \dots & a_{pm} \end{bmatrix} \begin{bmatrix} s_{i1} \\ \dots \\ s_{im} \end{bmatrix}$$
$$\boldsymbol{x}_i = \boldsymbol{A}\boldsymbol{s}_i. \tag{4.38}$$

Actually the more general case of unmixing considers a nonlinear transformation and an additional error term

$$\boldsymbol{x}_i = f(\boldsymbol{s}_i) + \epsilon_i. \tag{4.39}$$

The model (4.38) is the base for the blind source separation method. "Blind" means that little or nothing is known about the mixing matrix and we must make few assumptions on the sources s_i.

The principle behind the techniques of PCA and FA is to limit the number of components s_i to be small that is $m < p$ and is based on obtaining non-correlated sources.

ICA considers the case of $p = m$, obtains estimates of $\boldsymbol{A}$, and to finds components s_i that are independent as possible.

After estimating the matrix $\boldsymbol{A}$, we can compute its inverse, and obtain the independent component simply by:

$$\boldsymbol{s} = \boldsymbol{W}\boldsymbol{x} \tag{4.40}$$

Some of the characteristics of the procedure are:

- We fix the variances of the components to be one.
- We cannot determine the order of the components.
- To estimate $\boldsymbol{A}$ at most one of the components can be Gaussian. The independent components must be Gaussian.

The key to estimating the ICA model are non-Gaussianity and independence. There are several methods: maximum likelihood and network entropy, mutual information and Kulback-Leibler divergence, nonlinear cross correlations, nonlinear PCA, higher-order cumulants, and weighted covariance matrix, see Hyvärinen [94]. For a Bayesian approach, see Rowe [181].

Here, we outline the Infomax principle of network entropy maximization, which is equivalent to maximum likelihood (see Hyvärinen and Oja [96]). Here, we follow Stone and Porril [201] and Stone [200].

Consider Equation (4.40), $\boldsymbol{s} = \boldsymbol{W}\boldsymbol{x}$. If a signal or source s has a cumulative density function (cdf) g, then the distribution of $g(s)$ by the probability integral transformation has a uniform distribution and has maximum entropy.

The unmixing W can be found by maximizing the entropy of $H(U)$ of the joint distribution $U = (U_1, \ldots, Up) = (g_1(s_1), \ldots, g_p(s_p))$, where $s_i = \boldsymbol{W}\boldsymbol{x}_i$. The correct g_i has the same form as the cdf of the $\boldsymbol{x}_i$, which is sufficient to approximate these cdfs by sigmoid functions,

$$U_i = \tanh(s_i). \tag{4.41}$$

Given that $U = g(\boldsymbol{W}\boldsymbol{x})$, the entropy $\boldsymbol{H}(\boldsymbol{U})$ is related to $\boldsymbol{H}(x)$ by:

$$H(\boldsymbol{U}) = H(\boldsymbol{x}) + E\left(\log|\frac{\partial U}{\partial \boldsymbol{x}}|\right). \tag{4.42}$$

Given that we wish to find $\boldsymbol{W}$ that maximizes $\boldsymbol{H}(\boldsymbol{U})$, we can ignore $\boldsymbol{H}(\boldsymbol{x})$ in (4.42). Now,

$$\frac{\partial U}{\partial \boldsymbol{x}} = |\frac{\partial U}{\partial \boldsymbol{s}}||\frac{\partial \boldsymbol{s}}{\partial \boldsymbol{x}}| = \prod_{i=1}^{p} g'(s_i)|\boldsymbol{W}|. \tag{4.43}$$

Therefore, Equation (4.42) becomes

$$H(U) = H(\boldsymbol{x}) + E\left(\sum_{i=1}^{p} g(s_i')\right) + \log|\boldsymbol{W}|. \tag{4.44}$$

Given a sample of size n, the term $E\left(\sum_{i=1}^{p} g(s_i)\right)$ can be estimated by

$$E\left(\sum_{i=1}^{p} g(s_i)\right) \approx \frac{1}{n}\sum_{j=1}^{n}\sum_{i=1}^{p} \log g_i'\left(s_i^{(j)}\right). \qquad (4.45)$$

Ignoring $H(\boldsymbol{x})$, Equation (4.44) yields a new function that differs from $H(U)$ by $H(\boldsymbol{x})$.

$$h(\boldsymbol{W}) = \frac{1}{n}\sum_{j}^{n}\sum_{i}^{p} \log g_i'\left(s_i^*\right) + \log|\boldsymbol{W}|. \qquad (4.46)$$

Defining the cdf $g_i = \tanh$ and recalling that $g' = (1 - g^2)$ we obtain

$$h(\boldsymbol{W}) = \frac{1}{n}\sum_{j=1} n \sum_{i=1}^{p} \log(1 - s_i'^2) + \log|\boldsymbol{W}| \qquad (4.47)$$

whose maximization with respect to $\boldsymbol{W}$ yields

$$\triangledown \boldsymbol{W} = \frac{\partial H(\boldsymbol{W})}{\partial \boldsymbol{W}} = \frac{\partial h(W)}{\partial (W)} = (\boldsymbol{W}')^{-1} + 2.\boldsymbol{s}.\boldsymbol{x}', \qquad (4.48)$$

and an unmixing matrix can be found by taking small steps of size η to $\boldsymbol{W}$

$$\Delta \boldsymbol{W} = \eta\left(\boldsymbol{W}'^{-1} - 2.\boldsymbol{s}.\boldsymbol{x}'\right). \qquad (4.49)$$

After rescaling, it can be shown (see Lee [122, p. 41, p. 45]) that a more general expression is

$$\Delta \boldsymbol{W} \approx \begin{cases} (\boldsymbol{I} - \tanh(\boldsymbol{s})\boldsymbol{s}' - \boldsymbol{s}\boldsymbol{s}')\boldsymbol{W} & \text{super-Gaussian} \\ (\boldsymbol{I} + \tanh(\boldsymbol{s})\boldsymbol{s}' - \boldsymbol{s}\boldsymbol{s}')\boldsymbol{W} & \text{sub-Gaussian} \end{cases}. \qquad (4.50)$$

Super-gaussian random variables have typically a "spiky" probability density function (PDF) with heavy tails, i.e., the PDF is relatively large at zero and at large values of the variable, while being small for intermediate values. A typical example is the Laplace (or double-exponential) distribution. In contrast, sub-Gaussian random variables have typically a "flat" PDF that is rather constant near zero, and very small for large values of the variables. A typical example is uniform distribution.

We end this section with two illustrations of the applications of ICA and PCA to simulated examples from Lee [122, p. 32]. The first simulated example involves two uniformly distributed sources s_1 e s_2.

The sources are linearly mixed by:

$$\boldsymbol{x} = \boldsymbol{A}\boldsymbol{s} \qquad (4.51)$$

$$\begin{bmatrix} x_1 \\ x_2 \end{bmatrix} = \begin{bmatrix} 1 & 2 \\ 1 & 1 \end{bmatrix} \begin{bmatrix} s_1 \\ s_2 \end{bmatrix}.$$

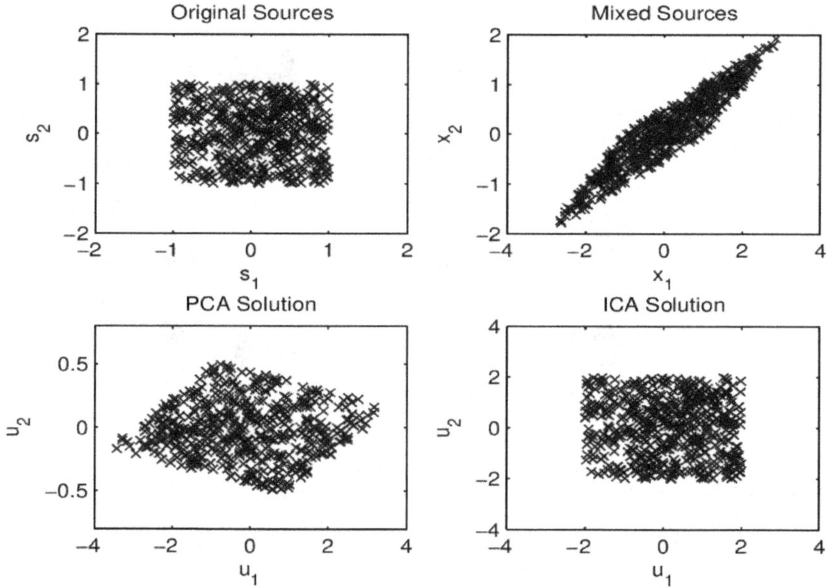

Figure 4.21
PCA and ICA transformation of uniform sources. Top-left: scatter plot of the original sources. Top-right: the mixtures. Bottom-left: the recovered sources using PCA. Bottom-right: the recovered sources using ICA.

Figure 4.21 shows the results of applying of the mixing transformation and the application of PCA and ICA. We see that PCA only spheres (decorrelates) the data, while ICA not only spheres the data but also rotates it such that $\hat{s}_1 = u_1$ and $\hat{s}_2 = u_2$ have the same directions of s_1 and s_2.

The second example in Figure 4.22 shows the time series of two speech signals s_1 and s_2. The signal is linearly mixed as the previous example in Equation (4.51). The solution indicates that the recovered sources are permutated and scaled.

4.2.3 PCA Networks

There are several neural network architectures for performing PCA (see Diamantaras and Kung [59] and Hertz et al. [87]). They can be:

- Autoassociator type. The network is trained to minimize

$$\sum_{i=1}^{n} \sum_{k=1}^{p} \left(y_{ik}(\boldsymbol{x} - x_{ik})\right)^2 \tag{4.52}$$

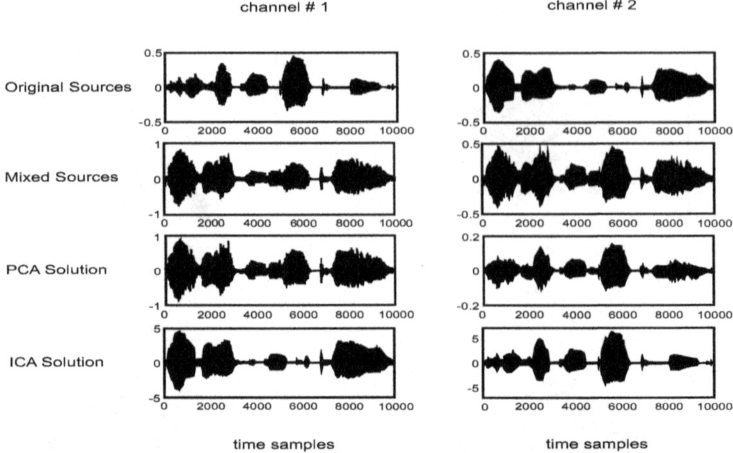

Figure 4.22
PCA and ICA transformation of speech signals (Top: the original sources, second row: the mixtures, third row: the recovered sources using PCA, bottom: the recovered sources using ICA).

or

$$||\boldsymbol{x}' - F_1(\boldsymbol{x})|| = ||A'(\boldsymbol{x} - \boldsymbol{\mu})||. \tag{4.53}$$

The architecture with two layer and identity activation function is shown in Figure 4.23.

- Networks based on Oja (Hebbian) rule

  The network is trained to find $\boldsymbol{W}$ that minimizes

$$\sum_{i=1}^{n} ||\boldsymbol{x}_i - \boldsymbol{W}'\boldsymbol{W}\boldsymbol{x}_i||^2 \tag{4.54}$$

  and its architecture is shown in Figures 4.24 and 4.25.

Several update rules have been suggested, all of which follow the same matrix equation

$$\Delta \boldsymbol{W}_l = \eta_l \boldsymbol{y}_l \boldsymbol{x}_l' - \boldsymbol{K}_l \boldsymbol{W}_l, \tag{4.55}$$

where $\boldsymbol{x}_l$ and $\boldsymbol{y}_l$ are the input and output vectors, η_l the learning rate and $\boldsymbol{W}_l$ is the weight matrix at the l-th iteration, K_l is a matrix whose choice depends on the specific algorithm, such as the following:

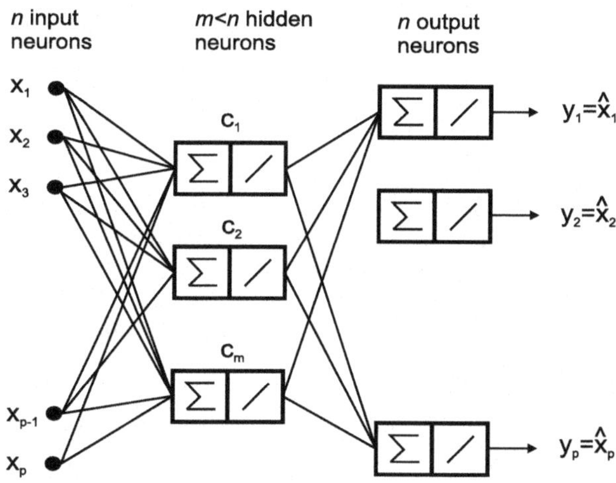

Figure 4.23
Autoassociative PCA networks.

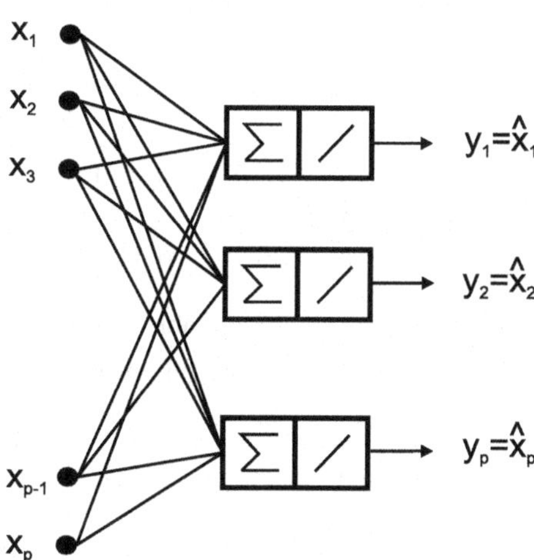

Figure 4.24
Oja rule PCA network architecture.

1. Williams rule: $\boldsymbol{K}_l = \boldsymbol{y}_l\boldsymbol{y}_l'$

2. Oja-Karhunen's rule: $\boldsymbol{K}_l = 3\boldsymbol{D}(\boldsymbol{y}_l\boldsymbol{y}_l') + 2L(\boldsymbol{y}_n\boldsymbol{y}_n')$

3. Sanger's rule: $\boldsymbol{K}_l = L(\boldsymbol{y}_l\boldsymbol{y}_l')$

where $D(\boldsymbol{A})$ is a diagonal matrix with diagonal entries equal to those of $\boldsymbol{A}$, $L(\boldsymbol{A})$ is a matrix whose entries above and including the main diagonal are zero. The other entries are the same as that of $\boldsymbol{A}$. For example:

$$D\begin{bmatrix}2 & 3\\ 3 & 5\end{bmatrix} = \begin{bmatrix}2 & 0\\ 0 & 5\end{bmatrix} \qquad L\begin{bmatrix}2 & 3\\ 3 & 5\end{bmatrix} = \begin{bmatrix}0 & 0\\ 3 & 0\end{bmatrix}. \tag{4.56}$$

To illustrate the computation involved, consider the $n = 4$ sample of a $p = 3$ variable data in Table 4.10.

Table 4.10
Data $p = 3$, $n = 4$.

n	x_1	x_2	x_3
1	1.3	3.2	3.7
2	1.4	2.8	4.1
3	1.5	3.1	4.6
4	1.1	3.0	4.8

The covariance matrix is

$$S^2 = \frac{1}{5}\begin{bmatrix}0.10 & 0.01 & -0.11\\ 0.01 & 0.10 & -0.10\\ -0.11 & -0.10 & 0.94\end{bmatrix}, \tag{4.57}$$

the eigenvalues and eigenvectors are shown in Table 4.11.

Now, we will compute the first principal component using a neural network architecture.

We choose $\eta_l = 1.0$ for all values of l, and select the initial values $w_1 = 0.3$, $w_2 = 0.4$ and $w_3 = 0.5$. For the first input $\boldsymbol{x}_1 = (0, 0.2, -0.7)$, we have

$$y = (0.3, 0, 4, 0, 5).(0, 0.2, -0.7) = -0.27, \tag{4.58}$$
$$\triangle \boldsymbol{W} = (-0.27(0, 0.2, -0.7) - (-0.27)^2.(0.3, 0.4, 0.5)), \tag{4.59}$$
$$\boldsymbol{W} = (0.3, 0.4, 0.5) + \triangle W = (0.278, 0.316, 0.652). \tag{4.60}$$

Table 4.11
SVD of the covariance matrix S^2.

eigenvalue		eigenvectors		
		$\boldsymbol{V}_1$	$\boldsymbol{V}_2$	$\boldsymbol{V}_3$
d_1	0.965	-0.823	0.553	-0.126
d_2	0.090	-0.542	-0.832	-0.115
d_3	0.084	-0.169	-0.026	-0.985

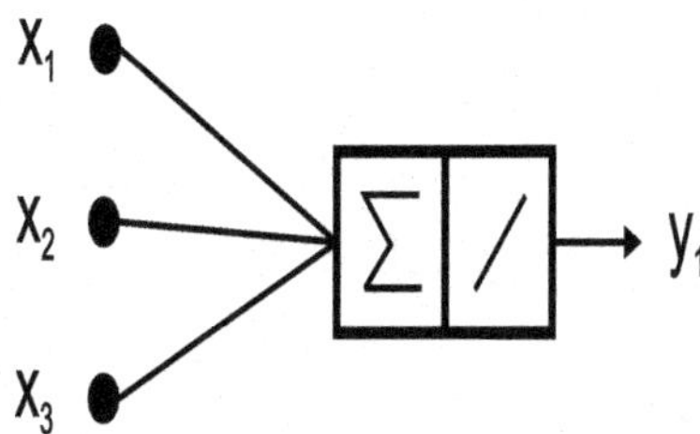

Figure 4.25
Oja rule.

For the next input $\boldsymbol{x}_2 = (0.1, -0.2, -0.3)$, we have

$$y = -0.231, \tag{4.61}$$

$$\triangle \boldsymbol{W} = (-0.231(0.10, -0.20, -0.30) - (-0.231)^2.(0.278, 0.316, 0.651)), \tag{4.62}$$

$$\boldsymbol{W} = (0.278, 0.316, 0.652) + \triangle W = (0.240, 0.346, 0.687). \tag{4.63}$$

Subsequent presentations of $\boldsymbol{x}_3, \boldsymbol{x}_4$, and $\boldsymbol{x}_5$ change the weight matrix to $(0.272, 0.351, 0.697)$, $(0.238, 0.313, 0.751)$ and $(0.172, 0.293, 0.804)$, respectively. This process is repeated, cycling through the input vectors $x_1, x_2, \ldots, x_5$. By the end of the second iteration, the weights become $(-0.008, -0.105, 0.989)$ and at the end of the third interaction, the weights change to $(-0.111, -0.028, 1.004)$. The weight adjustment process continues in this manner, resulting in the first principal component.

Further applications are mentioned in Diamantaras and Kung [59]. Most of the applications are in face, vowel, speech recognition. A comparison of the performance of the alternative rules is presented in Diamantaras and Kung [59] and Nicole [153]. The latter author also compares the alternative algorithms for neural computation of PCA with the classical batch SVD using Fisher's iris data [72].

4.2.4 Nonlinear PCA Networks

The nonlinear PCA network is trained to minimize:

$$\frac{1}{n}\sum_{i=1}^{n}||\boldsymbol{x}_i - \boldsymbol{W}'\varphi(\boldsymbol{W}\boldsymbol{x})||^2 \tag{4.64}$$

where φ is a nonlinear function with scalar arguments.

Since from Kolmogorov theorem, a function φ can be approximated by a sum of sinoidals, the architecture of the nonlinear PCA network is as in Figure 4.26.

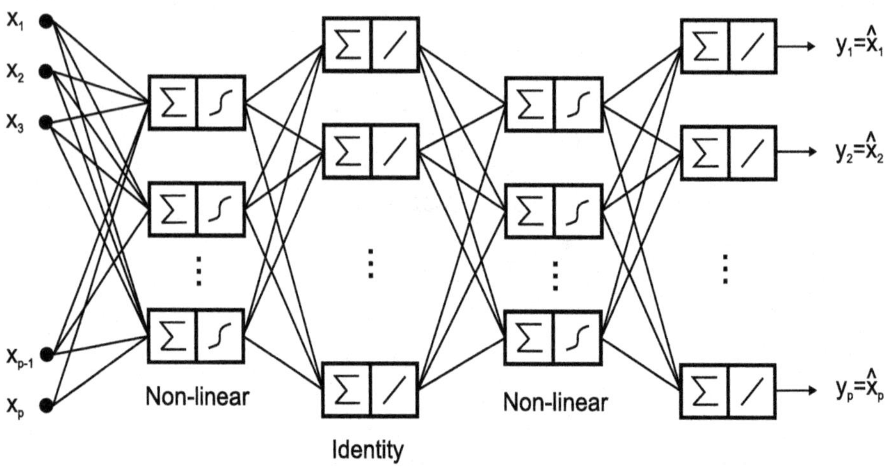

Figure 4.26
Nonlinear PCA network.

Application on face recognition is given in Diamantaras and Kung [59].

4.2.5 FA Networks

The application of artificial neural networks to factor analysis has been implemented using two types of architectures:

1) Using autoassociative networks

This network implemented by Sun and Lai [202] is called SunFA.

The concept consists in representing the correlation matrix $\boldsymbol{R} = (r_{ij})$ by cross-products (or outer product) of two factor symmetric matrices $\boldsymbol{F}_{p\,\mathrm{x}\,m}$ and $\boldsymbol{F}'_{p\,\mathrm{x}\,m}$.

Here the network architecture is that of Figure 4.23 and when there are m common factors the network is trained to minimize the function

$$\begin{aligned} E =&(r_{12} - \sum_{k=1}^{m} f_{1k} f_{2k})^2 + (r_{13} - \sum_{k=1}^{m} f_{1k} f_{3k})^2 + \dots \\ &+ (r_{1p} - \sum_{k=1}^{m} f_{1k} f_{pk})^2 + (r_{23} - \sum_{k=1}^{m} f_{2k} f_{3k})^2 + \dots \\ &(r_{2p} - \sum_{k=1}^{m} f_{2k} f_{pk})^2 + \dots + (r_{p-1,p} - \sum_{k=1}^{m} f_{p-1,k} f_{pk})^2. \end{aligned} \tag{4.65}$$

Sun and Lai [202] report the result of an application to measure ability of 12 students (verbal, numerical, etc.), and a simulation result comparing their method to four other methods (factor analysis, principal axes, maximum likelihood, and image analysis).

Five correlation matrices were used for samples of size $n = 82, 164$, and 328 with factors matrices $\boldsymbol{A}$, $\boldsymbol{B}$, and $\boldsymbol{C}$. Ten sets of correlation submatrices were sampled from each correlation matrix. They found that SunFA was the best method.

2) Using PCA (Hebbian) networks

Here a PCA network is used to extract the principal components of the correlation matrix and to choose the number of factors (components) m that will be used in the factor analysis. Then a second PCA network extracts m factors chosen in the previous analysis.

This procedure has been used by Delichere and Memmi [57], Delichere and Memmi [56] and Calvo [30]. Delichere and Memmi use an updating algorithm called the generalized Hebbian algorithm (GHA) and Calvo's adaptive principal component extraction (APEX). See Diamantaras and Kung [59].

We outline Calvo's application. The data consisted of school grades of eight students in four subjects (mathematics, biology, French, and Latin). The first PCA network shows that the first component explained 73% of the total variance and the second explained 27%.

The second network used the architecture shown in Figure 4.27.

Calvo obtained the results in Tables 4.12 and 4.13 and Figures 4.28 and 4.29, he also reported that the eigenvalues were equal within (± 0.01) to the results obtained on commercial software (SPPS).

4.2.6 Correspondence Analysis (CA) Networks

The only reference outlining how neural networks can be used for CA is Lebart [117]. The obvious architecture is the same for SVD or PCA shown in Figure 4.30. Here, we take

$$Z_{ij} = \frac{f_{ij} - f_{i\bullet} f_{\bullet j}}{\sqrt{f_{i\bullet} f_{\bullet j}}}. \tag{4.66}$$

Unfortunately, there does not seem to be any application on real data yet.

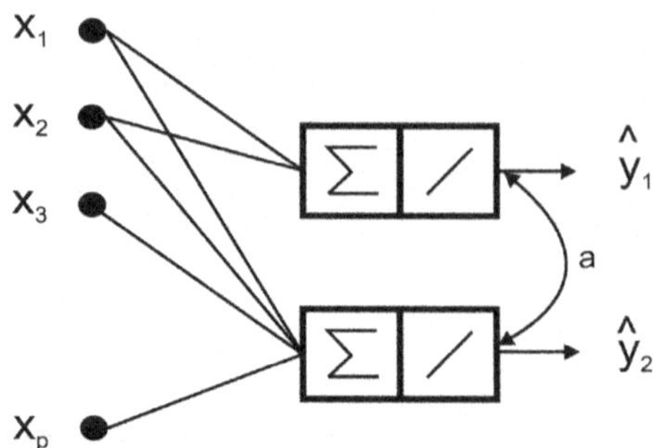

Figure 4.27
Factor analysis APEX network.

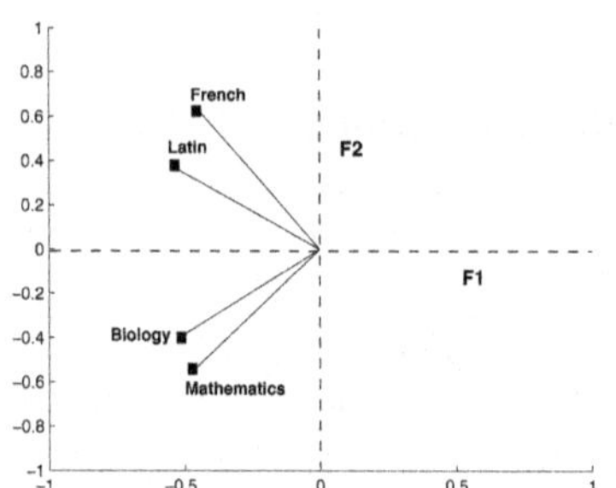

Figure 4.28
Original variables in factor space.

Figure 4.29
Students in factor space.

Table 4.12
School grades in four subjects.

Student	Math	Biology	French	Latin
1	13	12.5	8.5	9.5
2	14.5	14.5	15.5	15
3	5.5	7	14	11.5
4	14	14	12	12.5
5	11	10	5.5	7
6	8	8	8	8
7	6	7	11	9.5
8	6	6	5	5.5

Table 4.13
Loadings for the two factors retained.

	Factor 1	Factor 2
Mathematics	-0.4759	-0.5554
Biology	-0.5284	-0.4059
French	-0.4477	0.6298
Latin	-0.5431	0.3647

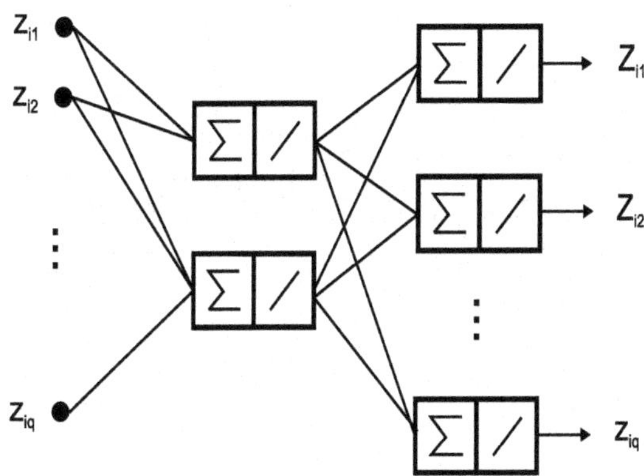

Figure 4.30
CA network.

4.2.7 Independent Component Analysis (ICA) Networks

The usual procedure in ICA follows the following steps:

- Standardize the data.
- Whiten (decorrelate, i.e., apply PCA).
- Obtain the independent components (e.g., Informax, i.e., NPCA).

One possible architecture is shown in Figure 4.31, where

$$\boldsymbol{x} = \boldsymbol{Q}\boldsymbol{s} + \boldsymbol{\varepsilon}$$

and the steps are:

- Whitening: $\boldsymbol{v}_k = \boldsymbol{V}\boldsymbol{x}_k$.
- Independent components: $y_k = \boldsymbol{W}v_k$.

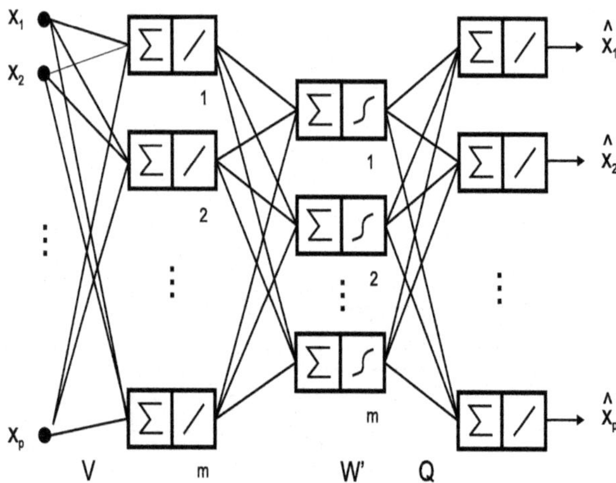

Figure 4.31
ICA neural network.

Now, we present some recent applications of ICA.

Fisher's iris data: Lee [122] compares classification algorithms with a classification based on ICA. The data set contains 3 classes where each class refers to a type of iris plant, 4 variables, and 50 observations in each class. The training set contained 75% of the observation and the testing set 25%. The classification error in the test data was 3%. A simple classification with boosting and a k-means clustering gave errors of 4,8% and 4,12% respectively. Figure 4.32 presents the independent directions (basis vectors).

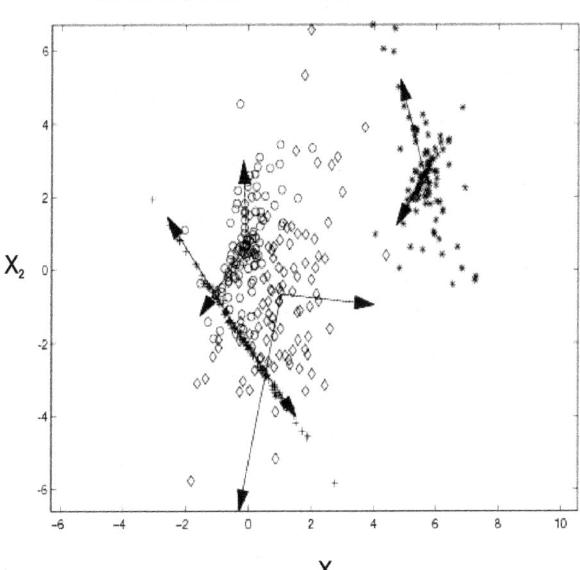

Figure 4.32
An example of classification of a mixture of independent components. There are 4 different classes, each generated by two independent variables and bias terms. The algorithm is able to find the independent directions (basis vectors) and bias terms for each class.

Draghici and Potter [61] implement a data mining technique based on the method of ICA to generate reliable independent data sets for different HIV therapies. They consider a coupled model that incorporates three algorithms:

i. ICA model is used for data refinement and normalization. The model accepts a mixture of CD4+, CD8+ and viral load data to a particular patient and normalizes it with broader data sets obtained from other HIV libraries. The ICA is used to isolate groups of independent patterns embedded in the considered libraries.

ii. Cluster (Kohonen Map) networks are used to select such patterns chosen by ICA algorithm that are close to the considered input data. These two mechanisms helped to select similar data and also to improve the system by throwing away unrelated data sets.

iii. Finally, a nonlinear regression model is used to predict future mutations in the CD4+, CD8+, and viral loads.

The authors point out a series of advantages of their method over the usual mathematical modelling using a set of simultaneous differential equations, which requires an expert to incorporate the dynamic behavior in the equations.

Also of interest are Karhunen et al. [107], Lee et al. [123], and Giannakopoulos et al. [77].

In the following, we discuss some illustrations of financial applications mentioned by Hyvärinen et al. [95]:

i. Application of ICA as a complementary tool to PCA in a study of a stock portfolio, allowing the underlying structure of the data to be more readily observed. This can be of help in minimizing the risk in investment strategies.

ii. ICA was applied to find the fundamental factor common to all stores that affects the cashflow, in the same retail chain. The effect of managerial actions could be analyzed.

iii. Time series prediction. ICA has been used to predict time series data by first predicting the common components and then transforming back to the original series.

Further applications in finance are found in Chan and Cha [35] and Cha and Chan [33]. The book of Girolami [78] also gives an application of clustering using ICA to the Swiss banknote data on forgeries.

4.3 Classification Networks

As mentioned before, two important tasks in pattern recognition are pattern classification and cluster or vector quantization. Here, we deal with the classification problem, whose task is to allocate an object characterized by a number of measurements into one of several distinct classes. Let an object (process, image, individual, etc.) be characterized by the k-dimensional vector $\mathbf{x} = (x_1, \ldots, x_k)$ and let $c_1, c_2, \ldots, c_L$ be labels representing each of the classes.

The objective is to construct an algorithm that will use the information contained in the components $x_1, \ldots, x_k$ of the vector $\mathbf{x}$ to allocate this vector to one of the L distinct categories. Typically, the pattern classifier or discrimination function calculates L functions (similarity measures), P_1, $P_2, \ldots, P_L$ and allocates the vectors $\mathbf{x}$ to the class with the highest similarity measures.

Many useful ideas and techniques have been proposed for more than 60 years of research in classification problems. Pioneer works are Fisher [72] and Rao [172]. Reviews and comparisons can be seen in Raudys [174], Asparoukhov and Krzanowski [11], Bose [25], and Michie et al. [143].

Neural network models that are used for classification have already been presented in previous sections. Figures 3.1(d) and 3.1(e) and Figures 4.33 through 4.36 present taxonomies and examples of neural classifiers.

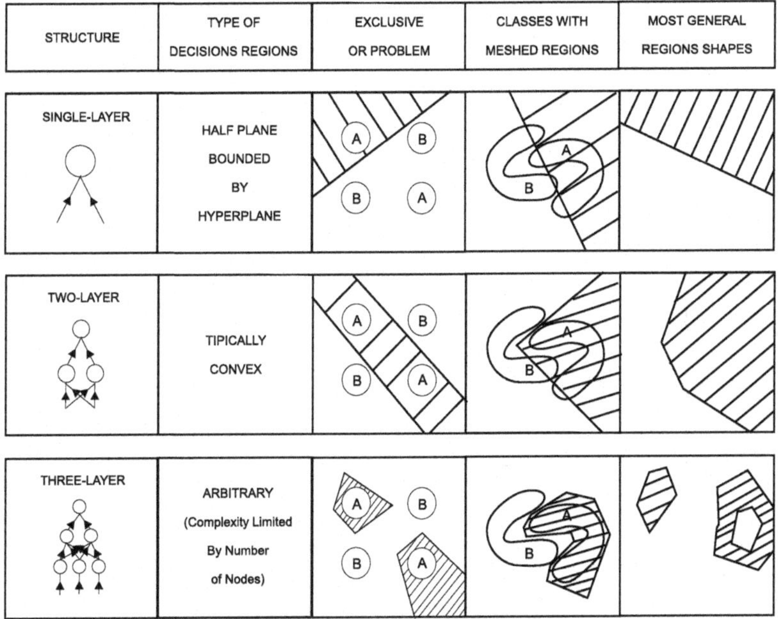

Figure 4.33
Taxonomy and decision regions. Types of decision regions that can be formed by single- and multi-layer perceptrons with hard limiting nonlinearities. Shading denotes regions for class A. Smooth closed contours bound input distributions.

Here we give some examples of classification problems solved using neural networks.

We start with an application of probabilistic neural networks (PNNs) in bankruptcy. Yang et al. [228] used data from 122 companies for the period 1984 to 1989 built on a bankrupt warning model for the US oil and gas industry. The data consist of five ratios to classify oil and gas companies into bankrupt and nonbankrupt groups. The five ratios are net cash flow to total assets, total debt to total assets, exploration expenses to total reserves, current abilities to total reserves, and the trend in total reserves calculated on the ratio of change from year to year. The first four rates are deflated. There are two separate data sets: one with deflated ratios and one without deflation.

The 122 companies are randomly divided into three sets: the training data set (33 nonbankrupt companies and 11 bankrupt companies), the validation data set (26 nonbankrupt companies and 14 bankrupt companies) and the testing data (30 nonbankrupt companies and 8 bankrupt companies).

Four methods of classification were tested: FISHER denotes Fisher discriminant analysis, FF denotes feedforward network, PNN means probabilistic neural net-

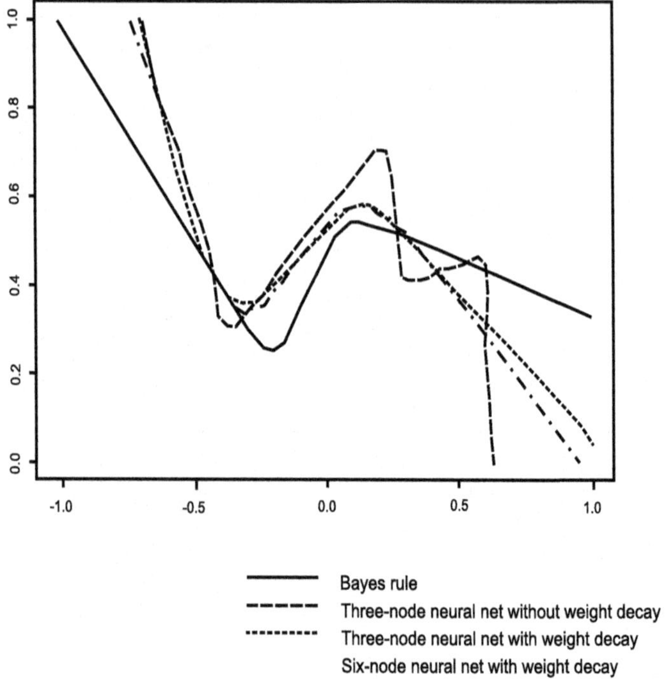

Figure 4.34
Boundaries for neural networks with tree and six hidden units (note how using weight decay smooths the fitted surface).

works, PNN* is probabilistic neural networks without data normalized. The normalization of data is in the form: $\mathbf{x}^* = \mathbf{x}/||\mathbf{x}||$.

Table 4.14 presents the results and we can see that the ranking of the classification model is: FISHER, BP, PNN, and PNN*.

Table 4.14
Number and percentage of correctly classified companies.

	Nondeflated data			Deflated data		
	Overall $n = 38$	Nonbankrupt $n = 30$	Bankrupt $n = 8$	Overall $n = 38$	Nonbankrupt $n = 30$	Bankrupt $n = 8$
FISHER	27 (71 %)	20 (67 %)	7 (88 %)	33 (87 %)	26 (87 %)	7 (88 %)
BP	28 (74 %)	24 (80 %)	4 (50 %)	30 (79 %)	30 (100 %)	0 (0 %)
PNN	25 (66 %)	24 (80 %)	1 (13 %)	26 (68 %)	24 (80 %)	2 (25 %)
PNN*	28 (74 %)	24 (80 %)	4 (50 %)	32 (84 %)	27 (90 %)	5 (63 %)

Bounds et al. [26] compared feedforward networks, radial basis functions, three

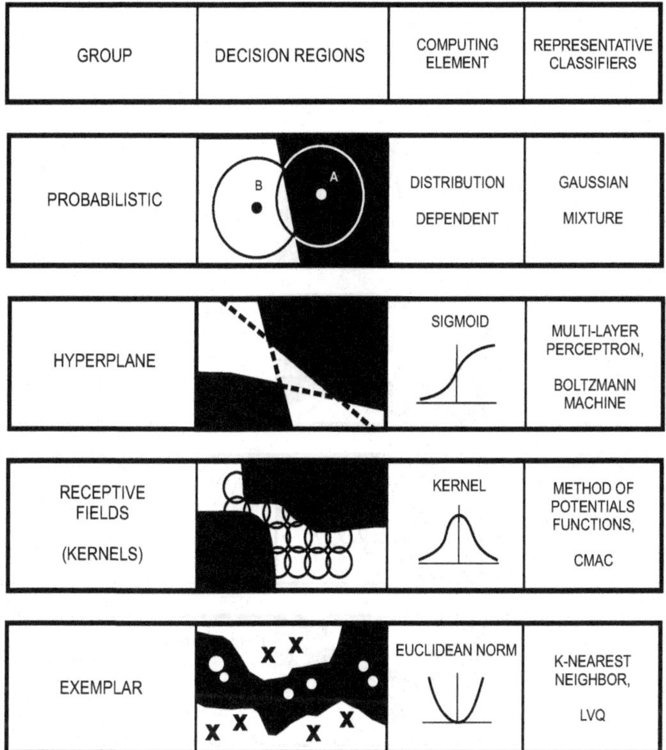

GROUP
DECISION REGIONS
COMPUTING ELEMENT
REPRESENTATIVE CLASSIFIERS
PROBABILISTIC
B
A
DISTRIBUTION DEPENDENT
GAUSSIAN MIXTURE
HYPERPLANE
SIGMOID
MULTI-LAYER PERCEPTRON,
BOLTZMANN MACHINE
RECEPTIVE FIELDS
(KERNELS)
KERNEL
METHOD OF POTENTIALS FUNCTIONS,
CMAC
EXEMPLAR
X X
X X X X
EUCLIDEAN NORM
K-NEAREST NEIGHBOR,
LVQ

Figure 4.35
Four basic classifier groups.

groups of clinicians and some statistical classification methods in the diagnostic of low back disorders. The data set referred to 200 patients with low back disorders.

Low back disorder was classified in four diagnostic categories: SLBP (simple low backpain), ROOTP (nerve root compression), SPARTH (spinal pathology, due to tumor, inflammation, or infection), AIB (abnormal illness behavior, with significant psychological overlay).

For each patient, the data was collected in a tick sheet that listed all the relevant clinical features. The patients' data were analyzed using two different subsets of the total data. The first data set (full assessment) contained all 145 tick sheet entries. This corresponds to all information including special investigations. The second data used a subset of 86 tick sheet entries for each patient. These entries correspond to a very limited set of symptoms that can be collected by paramedic personnel.

There were 50 patients for each of the four-diagnostic classes. Of the 50 patients in each class 25 were selected for the training data and the remaining for the test set.

The neural networks used were: feedforward network for individual diagnostic ($y = 1$) and ($y = 0$) for remaining three, feedforward with two outputs (for the four-

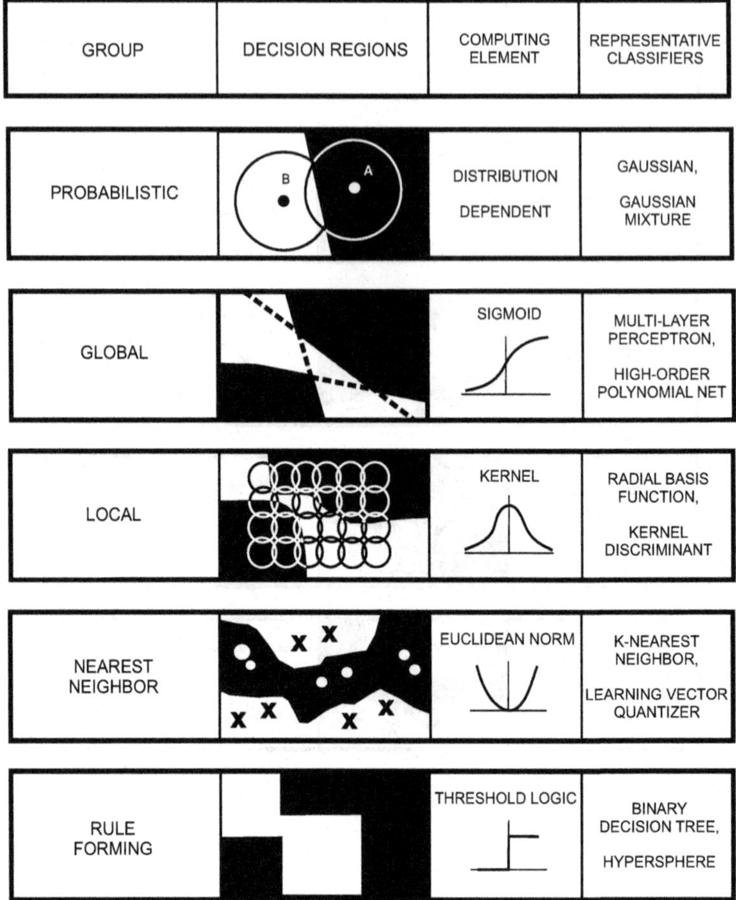

GROUP
DECISION REGIONS
COMPUTING ELEMENT
REPRESENTATIVE CLASSIFIERS
PROBABILISTIC
B
A
DISTRIBUTION
DEPENDENT
GAUSSIAN,
GAUSSIAN MIXTURE
GLOBAL
SIGMOID
MULTI-LAYER PERCEPTRON,
HIGH-ORDER POLYNOMIAL NET
LOCAL
KERNEL
RADIAL BASIS FUNCTION,
KERNEL DISCRIMINANT
NEAREST NEIGHBOR
EUCLIDEAN NORM
K-NEAREST NEIGHBOR,
LEARNING VECTOR QUANTIZER
RULE FORMING
THRESHOLD LOGIC
BINARY DECISION TREE,
HYPERSPHERE

Figure 4.36
A taxonomy of classifiers.

class diagnostic), a network of individual networks with two outputs, and the mean of 10 feedforward networks runs after training. This network had one hidden layer.

The radial basis networks were designed to recognize all four diagnostic classes. The effects of the number of centers, positions of the centers, and different choices of basis functions were all explored.

Two other classification methods and CCAD (computed-aided diagnosis) system were also tested in these data. A summary of the results is shown in Tables 4.15 and 4.16. The performance of the radial basis function, CCAD and MLP is quite good.

Santos et al. [185] and Santos et al. [186] used a feedforward network in the diagnosis of hepatitis A and pulmonary tuberculosis and compared the network to logistic regression. The hepatitis A data [185] encompass 2815 individuals from a

Table 4.15
Symptomatic assessment: Percentage of correct test diagnoses.

Classification Method	Simple low back pain	Root pain	Spinal pathology	Abnormal illness behavior	Mean
Clinicians:					
Bristol neurosurgeons	60	72	70	84	71
Glasgow orthopaedic surgeons	80	80	68	76	76
Bristol general practitioners	60	60	72	80	68
CCAD	60	92	80	96	82
MLP					
Best 50-20-2	52	92	88	100	83
Mean 50-20-2	43	91	87	98	80
Best 50-0-2	44	92	88	96	80
Mean 50-0-2	41	89	84	96	77
Radial basis functions	60	88	80	96	81
Closest class mean (CCM) (Euclidean)					
50 Inputs	56	92	88	96	63
86 Inputs	68	88	84	96	84
Closest class mean (CCM) (scalar product)					
50 Inputs (K=22)	16	96	44	96	63
86 Inputs (K=4)	92	84	60	84	80
K Nearest Neighbor (KNN) (Euclidean)					
50 Inputs (K=22)	88	84	76	84	83
86 Inputs (K=4)	92	80	84	80	84
K Nearest Neighbor (KNN) (scalar product)					
50 Inputs (K=3)	24	100	56	92	68
86 Inputs (K=6)	100	84	56	84	81

Table 4.16
Full assessment: Percentage of correct test diagnoses.

Classification Method	Simple low back pain	Root pain	Spinal pathology	Abnormal illness behavior	Mean
Clinicians:					
Bristol neurosurgeons	96	92	60	80	82
Glasgow orthopaedic surgeons	88	88	80	80	84
Bristol general practitioners	76	92	64	92	81
CCAD	100	92	80	88	90
MLP					
Best 50-20-2	76	96	92	96	90
Mean 50-20-2	63	90	87	95	83
Best 50-0-2	76	92	88	96	88
Mean 50-0-2	59	88	85	96	82
Radial basis functions	76	92	92	96	89
Closest class mean (CCM) (Euclidean)					
85 Inputs	100	92	72	88	88
145 Inputs	100	88	76	88	88
Closet class mean (CCM) (scalar product)					
85 Inputs (K=22)	12	92	48	100	63
145 Inputs (K=4)	100	72	68	80	80
K Nearest Neighbor (KNN) (Euclidean)					
85 Inputs (K=19)	100	84	80	80	86
145 Inputs (K=4)	96	88	80	80	86
K Nearest Neighbor (KNN) (scalar product)					
85 Inputs (K=19)	48	96	52	92	68
145 Inputs (K=4)	92	92	72	76	83

survey sample from regions in Rio de Janeiro state. The seroprevalence of hepatitis was 36.6. From the 66 collected variables in the study, seven variables reflect information on the individuals, housing environment, and socioeconomic factors, for a test sample of 762 randomly chosen individuals. Table 4.17 presents the results for the multilayer perceptron and a logistic regression model.

Table 4.17
Sensitivity (true positive) and specificity (true negative).

Errors	Model	
	Multilayer progression	Logistic regression
Sensitivity (%)	70	52
Specificity (%)	99	99

The pulmonary tuberculosis data [186] covered 136 patients of the Hospital Universitário (HUCFF) of the Universidade Federal do Rio de Janeiro. It included 23 variables and the test sample contained 45 patients. A comparison was made using the multilayer perceptron (MLP) and classification and regression tree (CART) models. Table 4.18 summarizes the results in the test data.

Table 4.18
Sensitivity and specificity.

Errors	Model	
	Multilayer perceptron	Classification and regression tree
Sensitivity (%)	73	40
Specificity (%)	67	70

A general comparison of classification methods including neural networks using medical binary data is given in Asparoukhov and Krzanowski [11]. Table 4.19 reproduces one of their three tables to indicate the methods used (they also have a similar table for large (15-17) and moderate (10) numbers of variables).

They suggest that:

- All the effective classifiers for large samples are traditional ones (IBM - independent binary model behavior, LDF - linear discriminating function, LLM - 2 order loglinear model).
- Most of the effective classifiers for footnote-size samples are non-traditional ones (MLP - multilayer perceptron, LVQ - learning vector quantization neural network, MIP - mixed integer programming).

Arminger et al. [9] compared the logistic discrimination regression tree and neural networks in the analysis of credit risk. The dependent variable is whether the credit is paid off or not. The predictor variables are sex, employment, marital status, and own car and own telephone. The result on the test data is shown in Table 4.20.

Table 4.19
Leave-one-out error rate (multiplied by 10^3 for 6 variables[1].

Discriminant procedure	Data set									
	Psychology		Pulmonary		Thrombosis		Epilepsy		Aneurysm	
	$=$	$\neq$	$=$	$\neq$	$=$	$\neq$	$=$	$\neq$	$=$	$\neq$
Linear classifiers										
IBM	239	283	147	139	228	284	168	186	23	21
LDF	244	223	147	139	272	275	147	147	53	46
LD	229	266	147	167	302	343	193	186	28	25
MIP		-		139		325		101		23
Quadratic classifiers										
Bahadur(2)	219	201	152	139	243	343	197	155	23	21
LLM (2)	206	207	151	144	250	304	153	116	23	21
QDF	215	207	-	-	265	314	153	178	-	-
Nearest neighbor classifiers										
kNN-Hills(L)	273(2)	266(1)	187(1)	185(1)	265(1)	294(1)	153(2)	217(1)	363(1)	306(1)
kNN-Hall(L)	230(2)	217(2)	166(1)	144(1)	257(1)	324(1)	102(2)	124(1)	23(1)	21(1)
Other non-parametric classifiers										
Kernel	247	234	166	144	243	363	143	124	23	21
Fourier	271	223	166	154	316	373	177	147	23	21
MLP(3)		207		139		284		132		25
LVQ(c)		190(6)		146(6)		226(6)		109(6)		37(4)

[1] = prior probabilities.

$\neq$ = class individual number divided by design set size.

kNN-Hills(L) and lNN-Hall(L) = order of procedure.

MLP(h) = hidden layer neuron number.

LVQ(c) = number of codebooks per class.

Table 4.20
Classification of credit data.

	MLP	LR	CART
Good credit	53	66	65
Bad credit	70	68	66
Overall	66	67	66

For the lender point of view MLP is more effective.

Desai et al. [58] explore the ability of feedforward network, mixture-of-expert networks, and linear discriminant analysis in building credit scoring models in a credit union environmental, using data from three credit unions, split randomly in 10 datasets. The data for each credit union contain 962, 918, and 853 observations. One third of each were used as test. The results for the generic models (all data) and the customized models for each credit union are presented in Tables 4.21 and 4.22 respectively.

Table 4.21
Generic models.

Data set	Percentage correctly classified							
	mlp_g		mmn_g		lda_g		lr_g	
	% total	% bad	% total	% bad	% total	% bad	% total	% bad
1	79.97	28.29	80.12	28.95	79.80	27.63	80.30	33.55
2	82.96	35.06	81.80	38.31	80.60	31.82	81.40	35.06
3	81.04	36.44	79.20	47.46	82.40	37.29	82.70	40.68
4	82.88	33.80	83.33	38.03	83.49	40.85	83.49	44.37
5	79.21	32.88	78.59	43.15	80.60	37.67	80.01	39.04
6	81.04	53.69	80.73	35.57	80.58	38.93	81.35	42.95
7	80.73	44.30	79.82	44.97	80.73	36.24	82.57	41.61
8	78.89	50.00	79.82	41.96	80.58	32.88	81.35	39.73
9	81.92	43.87	80.43	32.90	81.04	30.97	81.35	33.55
10	79.51	62.50	80.73	32.64	81.35	33.33	82.42	40.28
average	80.75	42.08	80.46	38.39	81.12	34.76	81.70	39.08
p-value[1]			0.191	0.197	0.98	0.035	0.83	0.19

[1] The p-values are for one-tailed paired t-test comparing mlp_g results with the other three methods.

The results indicate that the mixture-of-experts performs better in classifying poor credit risks.

West [218] investigates the credit scoring accuracy of five neural network models: multilayer perceptron (MLP), mixture-of-experts, radial basis function, learning vector quantization, and fuzzy adaptive resonance. The neural networks credit scoring models are tested using 10-fold cross validation with two real data sets. Results are compared with linear discriminant analysis, logistic regression, k-nearest neighbor, kernel density estimation, and decision trees.

Results show that feedforward network may not be the most accurate neural network model and that both mixture of experts and radial basis function should be considered for credit scoring applications. Logistic regression is the most accurate of the traditional methods.

Table 4.23 summarizes the results.

Markham and Ragsdale [140] present an approach to classification based on a combination of linear discrimination function (which the label Mahalanobis distance measure - MDM) and feedforward network. The characteristics of the data set are

Table 4.22
Customized models.

Data set	Sample	Percentage correctly classified					
		mlp_c		lda_c		lr_c	
		% total	% bad	% total	% bad	% total	% bad
Credit union L:							
1	14.88	83.33	32.00	83.30	16.00	82.70	24.00
2	17.26	87.50	31.03	82.70	20.69	81.00	34.48
3	17.26	86.90	41.38	83.90	31.03	82.10	37.93
4	22.02	81.55	37.84	84.52	32.43	82.74	37.84
5	18.45	82.14	29.03	77.40	16.13	78.00	38.71
6	17.26	83.33	41.38	81.55	27.59	82.74	41.38
7	21.43	80.36	16.67	78.50	08.33	80.95	30.56
8	17.86	79.76	56.67	82.14	10.00	81.55	23.33
9	16.67	85.71	32.14	86.31	32.14	86.90	39.29
10	18.45	83.33	29.03	79.76	19.35	80.36	29.03
Credit union M:							
1	26.77	85.43	79.71	86.60	70.97	87.40	73.53
2	26.77	88.58	76.47	85.40	18.64	88.20	79.41
3	20.47	85.43	80.77	87.40	31.58	85.80	75.00
4	23.63	90.94	75.00	90.20	35.14	89.00	75.00
5	24.02	89.37	88.52	89.90	22.22	86.60	81.97
6	26.38	88.58	74.63	88.19	12.96	86.22	71.64
7	26.77	85.43	74.63	85.04	28.30	86.22	77.61
8	25.59	88.19	75.38	86.61	28.89	87.40	67.69
9	27.59	87.40	72.86	85.43	31.37	86.61	67.14
10	24.41	90.55	82.26	89.37	21.05	89.37	80.65
Credit union N:							
1	25.43	75.86	49.15	75.40	18.64	78.50	27.11
2	24.57	77.15	40.35	···	31.58	75.50	24.56
3	15.95	81.90	35.14	83.20	35.14	84.10	35.14
4	19.40	77.15	44.44	78.90	22.22	80.60	28.89
5	23.28	76.72	24.07	75.40	12.96	75.90	18.52
6	22.84	79.31	35.85	75.00	28.30	76.22	26.42
7	19.40	81.90	31.11	80.60	28.89	81.03	28.89
8	21.98	74.13	37.25	74.57	31.37	75.43	31.37
9	24.57	80.17	47.37	77.59	21.05	78.88	21.05
10	21.98	77.59	19.61	77.16	21.57	76.72	19.61
average		83.19	49.72	82.35	38.49	82.67	44.93
p-value[1]				0.018	5.7E-7	0.109	0.0007

··· = Missing values.

[1] The p-values are for one-tailed paired t-test comparing mlp_c results with the other two methods.

shown in Table 4.24. The process of splitting the data in validation and test set was repeated 30 times. The results are shown in Table 4.25.

Finally, a recent reference on comparisons of classification models is Bose [25]. Further results on classification related to neural networks can be seen in Lippmann [127, 128, 129], Ripley [176], and [140].

Table 4.23
Credit scoring errors of neural network models.

	German credit data[b]			Australian credit data[b]		
	Good credit	Poor credit	Overall	Good credit	Poor credit	Overall
Neural models[a]						
MOE	0.1428	0.4775	0.2434	0.1457	0.1246	0.1332
RBF	0.1347	0.5299	0.2540	0.1315	0.1274	0.1286
MLP	0.1352	0.5753	0.2672	0.1540	0.1326	0.1416
LVQ	0.2493	0.4814	0.3163	0.1710	0.1713	0.1703
FAR	0.4039	0.4883	0.4277	0.2566	0.2388	0.2461
Parametric models						
Linear discriminant	0.2771	0.2667	0.2740	0.0782	0.1906	0.1404
Logistic regression	0.1186	0.5133	0.2370	0.1107	0.1409	0.1275
Non-parametric models						
K nearest neighbor	0.2257	0.5533	0.3240	0.1531	0.1332	0.1420
Kernel density	0.1557	0.6300	0.3080	0.1857	0.1514	0.1666
CART	0.2063	0.5457	0.3044	0.1922	0.1201	0.1562

[a] Neural network results are averages of 10 repetitions.

[b] Reported results are group error rates averaged across 10 independent holdout samples.

Table 4.24
Data Characteristics.

Dataset	1	2
Problem domain	Oil quality rating	Bank failure prediction
Number of groups	3	2
Number of observations	56	162
Number of variables	5	19

Table 4.25
Comparison of classification methods on three-group oil quality and two-group bank failure data set.

	Percentage of Misclassified Observations in the Validation Sample					
	Three-group oil quality			Two-group bank failure		
Sample	MDM	NN1	NN2	MDM	NN1	NN2
1	22.22	18.52	3.70	10.00	3.75	2.50
2	14.81	18.52	11.11	17.50	10.00	7.50
3	11.11	11.11	0	20.00	11.25	7.50
4	11.11	0	0	15.00	10.00	5.00
5	18.52	22.22	14.81	11.25	6.25	3.75
6	3.70	0	0	11.25	5.00	1.25
7	14.81	0	0	12.50	8.75	5.00
8	29.63	22.22	11.11	13.75	7.50	3.75
9	7.41	0	0	12.50	6.25	5.00
10	18.52	7.41	0	12.50	7.50	2.50
11	3.70	0	0	17.50	8.75	6.25
12	18.52	11.11	0	20.00	11.25	7.50
13	7.41	0	0	15.00	6.25	3.75
14	7.41	0	0	13.75	5.00	5.00
15	11.11	0	0	16.25	6.25	2.50
16	7.41	0	0	18.75	10.00	6.25
17	18.52	7.41	3.70	25.00	13.75	8.75
18	11.11	3.70	0	13.75	10.00	6.25
19	3.70	0	0	16.25	8.75	5.00
20	14.81	7.41	0	13.75	6.25	3.75
21	11.11	0	0	15.00	7.50	3.75
22	14.81	3.70	0	13.75	6.25	2.50
23	25.93	18.52	11.11	15.00	8.75	5.00
24	22.22	11.11	11.11	20.00	10.00	6.25
25	18.52	7.41	0	15.00	11.25	5.00
26	14.81	7.41	0	17.50	8.75	5.00
27	7.41	0	0	18.75	8.75	7.50
28	14.81	3.70	3.70	16.25	5.00	2.50
29	0	0	0	15.00	6.25	3.75
30	14.81	7.41	3.70	16.25	7.50	5.00
Average	13.33	6.42	2.47	15.63	8.08	4.83

4.4 Multivariate Statistics Neural Network Models with Python

In this section, we use SOM for clustering and fitting data with the library NeuPy. One can use other libraries; for instance, PyMVPA[1], which stands for Python Multi-Variate Pattern Analysis, see Hanke et al. [84].

[1] http://www.pymvpa.org/

4.4.1 Clustering

The library scikit-learn provides a handwritten digit dataset that was initially collected by the National Institute of Standards and Technology (NIST). Each number is represented in a matrix of 8 by 8 pixels. Using SOM, we can organize the whole dataset without the network learning the meaning of the numbers. Program 4.1 organizes the numbers using the library NeuPy, in particular, its command `SOFM` from the self-organizing feature map implementation of the SOM algorithm.

Program 4.1
Clustering digits with SOM.

```
1  import numpy as np
2  import matplotlib.pyplot as plt
3  from matplotlib import gridspec
4  from neupy import algorithms
5
6  print("Loading the dataset...")
7  from sklearn import datasets
8  dataset = datasets.load_digits()
9  digits = dataset.data
10
11 print("Training in a grid...")
12 sofm = algorithms.SOFM(
13     n_inputs = 64,
14     learning_radius = 2,
15     features_grid = (10, 10),
16     shuffle_data = True,
17 )
18 sofm.train(digits, epochs = 20)
19 clusters = sofm.predict(digits).argmax(axis = 1)
20
21 print("Gathering the clustered digits...")
22 plt.figure(figsize = (8, 8))
23 grid = gridspec.GridSpec(10, 10)
24 for position in range(100):
25     if len(digits[clusters == position]) == 0:
26         d = np.zeros(64)
27     else:
28         d = digits[clusters == position][0]
29     plt.subplot(grid[position])
30     plt.imshow(d.reshape((8,8)), cmap = 'Greys')
31     plt.axis('off')
32 plt.show()
```

The program is divided in three parts, namely, loading the dataset from lines 6 to 9, training from lines 11 to 19, and gathering the clustered digits from lines 21

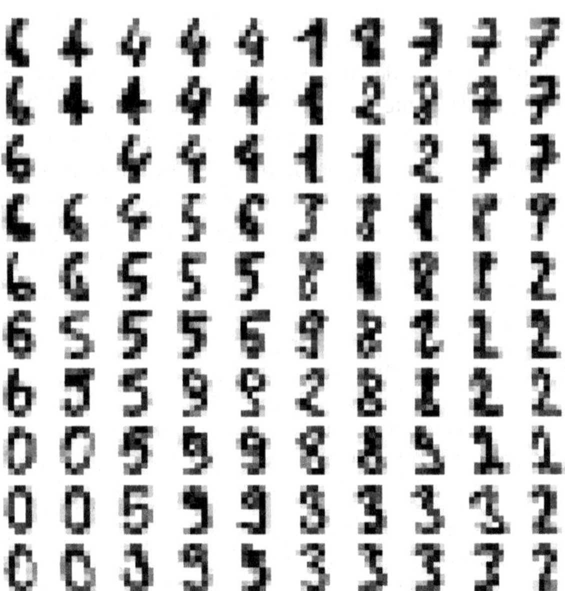

Figure 4.37
Cluster of digits after 20 epochs.

to 32. We set a grid 10 by 10 for training in line 15. Thus, we present 100 digits in a grid with the same dimensions set in line 23. The input in line 13 represents the image in a matrix of 8 by 8 pixels. Notice that the vector clusters might not have all positions of the numbers. In this case, we cannot recover the number because there is no position associated to the number. Thus, we need to verify if the position is in the vector clusters. Line 25 verifies if there is a position in clusters, and line 26 inserts white space if not. Otherwise, line 28 can recover the number in the position of the clusters.

Figure 4.37 depicts an output from Program 4.1 after 20 epochs. Notice that the 22nd position was not found in the clusters. Therefore, the line 26 filled the 22nd position with space.

We also use a dataset known as Fashion-MNIST, which has a name derived from the well-known MNIST (Modified National Institute of Standards and Technology) database. Instead of handwriting, Fashion-MNIST contains 60,000 clothing images for training and 10,000 for testing. Each image is in gray scale with 28 per 28 pixels. Each image is 28 by 28 pixels in grayscale. For more details about the dataset, see Xiao et al. [227]. The goal is to organize the clothes putting each article of clothing in a cluster. However, we do not know how many clusters are necessary. A solution is given in the Program 4.2, which might take several minutes to load the dataset from the internet and to train. The last and shortest step is to gather the clustered clothes into an image. Instead of using the whole dataset, we select only 600 images in the line 9. The SOM algorithm receives images in a vector of $784 = 28 \times 28$ pixels and organizes the clothes in a grid 20×30. The line 17 ensures that the clothes are not organized. The elements of the vector cluster contain positions for clothes in the clusters. Some clothing might not be clustered. Thus, we fill the space with a white image in line 28. Besides changing the number of epochs, we could uncomment the line 18 removing the symbol # for seeing details about the training process.

Program 4.2
Clustering with SOM.

```
1  import numpy as np
2  import matplotlib.pyplot as plt
3  from matplotlib import gridspec
4  from neupy import algorithms
5
6  print("Loading the dataset...")
7  from sklearn.datasets import fetch_openml
8  dataset = fetch_openml(name = "Fashion-MNIST")
9  clothes = dataset.data[:600]
10
11 print("Training in a grid...")
12 sofm = algorithms.SOFM(
13     n_inputs = 784,
14     learning_radius = 10,
15     features_grid = (20, 30),
16     reduce_radius_after = 10,
17     shuffle_data = True,
18 #    verbose = True
19 )
20 sofm.train(clothes, epochs = 2)
21 clusters = sofm.predict(clothes).argmax(axis = 1)
22
23 print("Gathering the clustered clothes...")
24 plt.figure(figsize = (15, 10))
25 grid = gridspec.GridSpec(20, 30)
26 for position in range(600):
27     if len(clothes[clusters == position]) == 0:
28         clothing = np.zeros(784)
29     else:
30         clothing = clothes[clusters == position][0]
31     plt.subplot(grid[position])
32     plt.imshow(clothing.reshape((28,28)), cmap = 'Greys')
33     plt.axis('off')
34 plt.show()
```

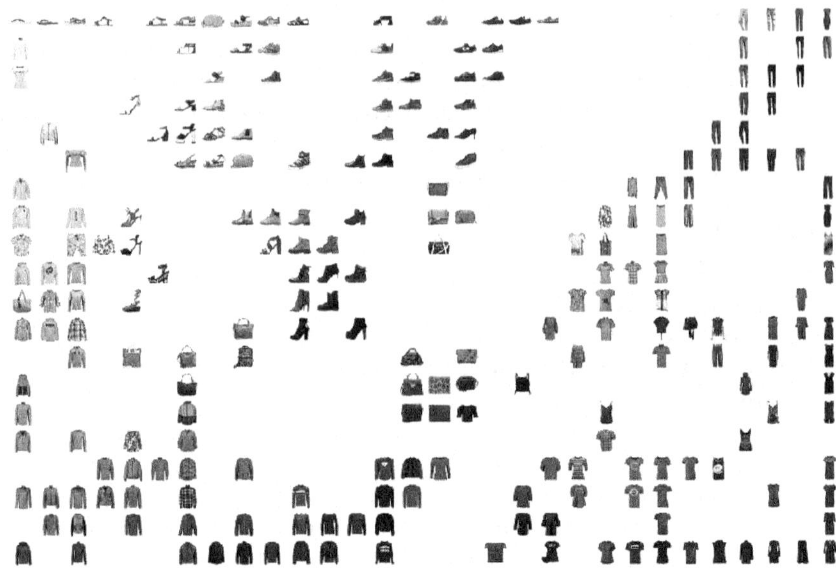

Figure 4.38
Cluster of clothes after 20 epochs.

Figure 4.38 depicts an output from Program 4.2 with the epoch set as 20. Figure 4.39 depicts an output with the epoch set as 200. Comparing, we see that the SOM organized more clothes over the time.

We can speed up using PCA to reduce from 784 to 400 components. To run Program 4.2 and its version with PCA, we can use functions in Python. Specifically, we can create the functions `Gathering()` and `Training(C,I)`, where the first argument `C` is the set of clothes for the case without PCA and for the case with PCA where the clothes were transformed by the PCA, the second argument `I` is the number of components. In this way, we can compare the elapsed time in seconds for processing the training with and without PCA. Program 4.3 shows us that the version of the training with PCA is approximately five times faster for 200 epochs.

Program 4.3
Using SOM for clustering with and without PCA.

```
import numpy as np
import matplotlib.pyplot as plt
from matplotlib import gridspec
from neupy import algorithms

def Gathering():
    print("Gathering the clustered clothes...")
```

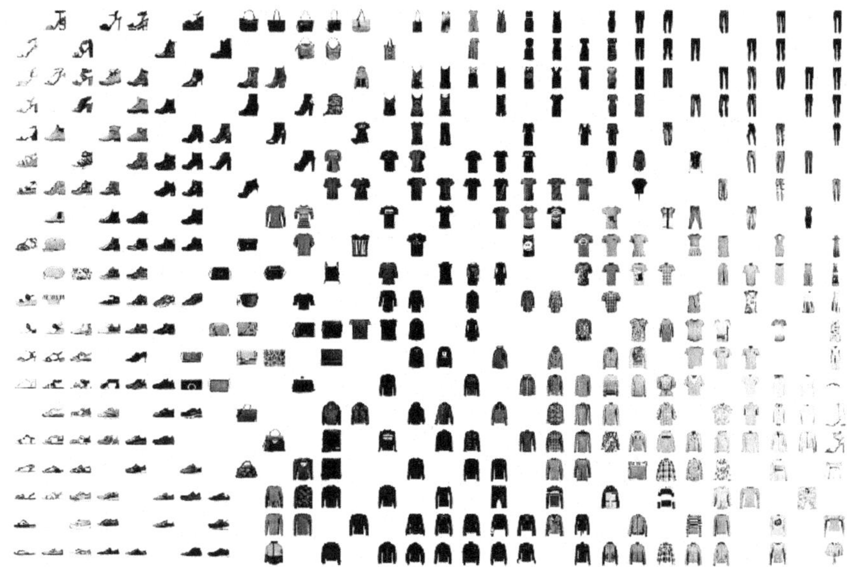

Figure 4.39
Cluster of clothes after 200 epochs.

```
        plt.figure(figsize = (15, 10))
        grid = gridspec.GridSpec(20, 30)
        for position in range(600):
            if len(clothes[clusters == position]) == 0:
                clothing = np.zeros(784)
            else:
                clothing = clothes[clusters == position
                    ↪ ][0]
            plt.subplot(grid[position])
            plt.imshow(clothing.reshape((28,28)), cmap =
                ↪ 'Greys')
            plt.axis('off')
        plt.show()

    def Training(C,I):
        print("Training in a grid...")
        sofm = algorithms.SOFM(
            n_inputs = I,
            learning_radius = 10,
            features_grid = (20, 30),
            reduce_radius_after = 10,
            shuffle_data = True,
```

```
        )
    sofm.train(C, epochs = 2)
    clusters = sofm.predict(C).argmax(axis = 1)
    return (clusters)


print("Loading the dataset...")
from sklearn.datasets import fetch_openml
dataset = fetch_openml(name = "Fashion-MNIST")
clothes = dataset.data[:600]

import time
start = time.process_time()
clusters = Training(clothes,784)
end = time.process_time()
sec = end - start

Gathering()

start = time.process_time()
print("PCA...")
from sklearn.decomposition import PCA
transformer = PCA(n_components = 400,random_state = 0)
clothes_transformed = transformer.fit_transform(clothes)

clusters = Training(clothes_transformed,400)

end = time.process_time()
sec_PCA = end - start

Gathering()

print("Elpsed time without PCA:", end = " ")
print(sec)
print("Elpsed time with    PCA:", end = " ")
print(sec_PCA)
```

4.4.2 Fitting Data

Similar to Program 3.2, we can use SOM for fitting the curve $\sin(x) + x$. We list the implementation in Program 4.4. Lines 7 and 8 organize the coordinates of points for line 17 to train the network with such points. Line 10 plots the function, and line 20 marks the points from the SOM on the graphic presented in Figure 4.40.

Program 4.4
Approximation of $\sin(x) + x$ using SOM.

```
1  import numpy as np
2  import matplotlib.pyplot as plt
3  from neupy import algorithms
4
5  x = np.arange(0, 10, .1)
6  y = (np.sin(x)+x).ravel()
7  v=np.concatenate((x,y)).reshape(2,100)
8  v=np.transpose(v)
9
10 plt.scatter(x, y, c='gray', marker=".")
11
12 sofm = algorithms.SOFM(
13       n_inputs=2,
14       features_grid=(1, 100),
15       learning_radius=2,
16 )
17 sofm.train(v, epochs=20)
18
19 weights = sofm.weight.reshape((2, 1, 100))
20 plt.scatter(weights[0,:,:],weights[1,:,:], marker="D
      ↪ ", color='black')
21
22 plt.xlim(0, 10)
23 plt.ylim(0, 10)
24
25 plt.show()
```

Instead of functions, we use a dataset as an input for fitting its points with SOM. The command `db.make_moons` provides the dataset in Program 4.5. Line 14 presents the 80 points from the dataset, and lines 15 and 16 present the points found by the network.

Program 4.5
Fitting points with SOM.

```
1  import matplotlib.pyplot as plt
2  from sklearn import datasets as db
3  from neupy import algorithms
4
5  sample, targets = db.make_moons(n_samples=80,
      ↪ random_state=0)
```

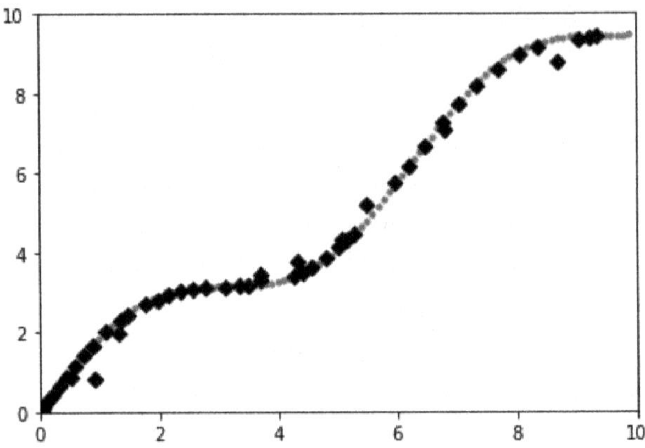

Figure 4.40
Approximation of $\sin(x) + x$ using SOM.

```

sofm = algorithms.SOFM(
        n_inputs=2,
        features_grid=(1, 80),
        learning_radius=2,
)
sofm.train(sample, epochs=100)

plt.scatter(sample[:,0],sample[:,1],  marker="o",
    ↪ color='#AAAAAA')
weights = sofm.weight.reshape((2, 1, 80))
plt.scatter(weights[0,:,:],weights[1,:,:], marker="D
    ↪ ", color='black')
plt.show()
```

Figure 4.41 depicts two possible outputs from Program 4.5: one with the 20 epoch set as in Figure 4.41(a) and another with the 200 epoch set as in Figure 4.41(b).

Note that SOM does not fit well based on a comparison of Figures 4.40 and 3.19.

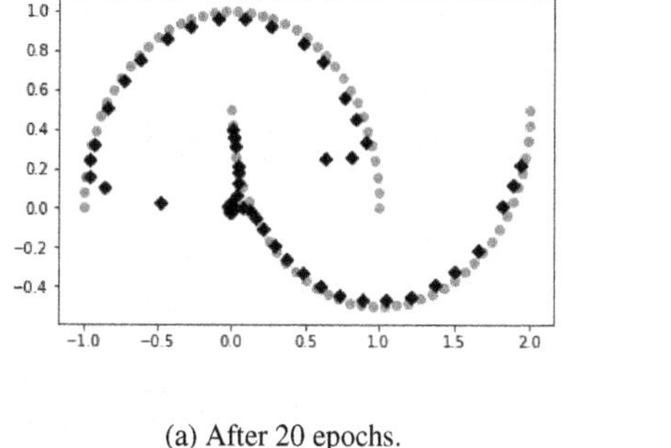

(a) After 20 epochs.

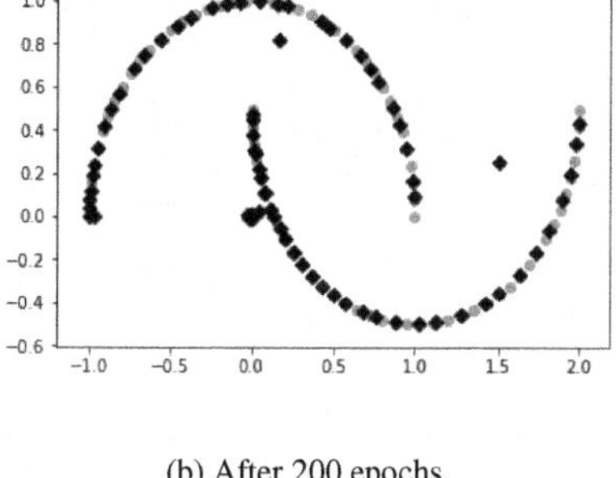

(b) After 200 epochs.

Figure 4.41
Fitting data using SOM.

5

Regression Neural Network Models

Regression models are studied in this chapter. On the one hand, the chapter presents the parametric: generalized linear model networks (GLIMN), logistic regression networks, and regression networks. On the other hand, it also presents nonparametric regression and classification networks, probabilistic neural networks, general regression neural networks, generalized additive model networks, regression and classification tree networks, and projection pursuit and feedforward networks.

5.1 Generalized Linear Model Networks (GLIMNs)

Generalized (iterative) linear models (GLIM) encompass many of the statistical methods most commonly employed in data analysis (see Nelder and McCullagh [150]). Beyond linear models with normal distributed errors, generalized linear models include logit and probit models for binary response variable and log-linear (Poisson-regression) models for counts.

A generalized linear model consists of three components:

1. A random component, in the form of a response variable Y with distribution from the exponential family which includes: the normal, Poisson, binomial, gamma, inverse-normal, negative-binomial distributions, etc.

2. A linear predictor
$$\eta_i = \alpha_0 + \alpha_1 x_{i1} + \ldots + \alpha_k x_{ik} \tag{5.1}$$
on which y_i depends. The x' are independent (covariates, predictors) variables.

3. A link function $L(\mu_i)$ that relates the expectation $\mu_i = E(Y_i)$ to the linear predictor η_i
$$L(\mu_i) = \eta_i. \tag{5.2}$$

The exponential family of distribution includes many common distributions and its parameters have some nice statistical properties related to them (sufficiency, attainment of Cramer-Rao bound). Members of this family can be expressed in the general form

$$f(y, \theta, \phi) = \exp\left(\frac{y\theta - b(\theta)}{\alpha(\phi) + c(y, \phi)}\right). \tag{5.3}$$

Table 5.1
Generalized linear models: mean, variance, and deviance functions.

	η	$V(\mu)$	$D(\hat{\mu})$
Normal	μ	1	$\sum(y-\hat{\mu})^2$
Binomial	$\ln[\mu/(1-\mu)]$	$\mu(1-\mu)$	$2\Sigma\{[y\ln(y/\hat{\mu})]+$ $+(n-y)\ln[(n-y)/(n-\hat{\mu})]\}$
Poisson	$\log\mu$	μ	$2\Sigma[y\ln(y/\hat{\mu})-(y-\hat{\mu})]$
Gamma	μ^{-1}	μ^2	$2\Sigma[-\ln(y/\hat{\mu})-(y-\hat{\mu})/\hat{\mu}]$

If ϕ is known, then θ is called the *natural* or *canonical* parameter. When $\alpha(\phi)=\phi$, ϕ is called the *dispersion* or *scale* parameter. It can be shown that

$$E(Y)=b'(\theta)=\mu \text{ and } \operatorname{Var}(Y)=\alpha(\phi)b''(\theta)=\alpha(\phi)\vee(\theta). \tag{5.4}$$

Fitting a model may be regarded as a way of replacing a set of data values $\boldsymbol{y}=(y_i\ldots y_n)$ by a set of fitted values $\boldsymbol{\mu}=(\hat{\mu}_1,\ldots,\hat{\mu}_n)$ derived from a model involving (usually) a relatively small number of parameters. One measure of discrepancy most frequently used is that formed by the logarithm of the ratio of likelihoods, called deviance (D).

Table 5.1 gives some usual choices of link functions for some distributions of $\boldsymbol{y}$ for the canonical link $\theta=\eta$.

The neural network architecture equivalent to the GLIM model is a perceptron with the predictors as inputs and one output. There are no hidden layers and the activation function is chosen to coincide with the inverse of the link function (L). This network is shown in Figure 5.1.

In what follows, we will deal with some particular cases: the logit regression and the regression models neural networks.

5.1.1 Logistic Regression Networks

The logistic model deals with a situation in which we observe a binary response variable y and k covariates x's. The logistic regression model is a generalized linear model with binomial distribution for $y_1\ldots y_n$ and logit link function. The equation

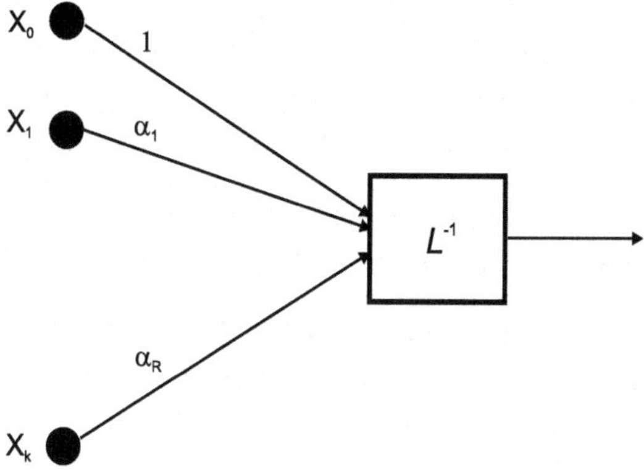

Figure 5.1
GLIM network.

takes the form

$$p = P(Y = 1/\underset{\sim}{x}) = \frac{1}{1 + \exp(-\beta_0 - \sum_{i=1}^{k} \beta_i x_i)} = \wedge \left(\beta_0 + \sum_{i=1}^{k} \beta_i x_i \right), \quad (5.5)$$

with $\wedge$ denoting the logistic function. The model can be expressed in terms of log odds of observing 1 given covariates x as

$$\log tp = \log \frac{P}{1 - \beta} = \beta_0 + \sum_{i=1}^{k} \beta_i x_i. \quad (5.6)$$

A natural extension of the linear logistic regression model is to include quadratic terms and multiplicative interaction terms

$$P = \Lambda \left(\beta_0 + \sum_{i=1}^{k} \beta_i x_i + \sum_{i=1}^{k} \gamma_i x_i^2 + \sum_{i<j} \delta_{ij} x_i x_j \right). \quad (5.7)$$

Other approaches to model the probability p include generalized additive models (Section 5.2.3) and feedforward neural networks.

For the linear logistic model, Equation (5.6) the neural network architecture is as in Figure 5.1 with the sigmoid link k function as activation function of the output neuron.

A feedforward network with hidden layer with r neurons and sigmoid activation function would have the form

$$P = \Lambda \left(w_0 + \sum_{j=1}^{r} w_j \Lambda(w_{oj} + \sum_{j=1}^{k} w_{ij} x_i) \right). \quad (5.8)$$

This representation of neural networks is the basis for analytical comparison of feedforward networks and logistic regression. For a neural network without a hidden layer and a sigmoid activation function, Equation (5.8) reduces to (5.6) and this network is called logistic perceptron in the neural network literature. Therefore, it does not make sense to compare a neural network with hidden layer and linear logistic regression. For a discussion of this and related measures, and other comparisons among these two models, see Schwarzer et al. [190] and Tu [211]. An attempt to improve the comparison and to use logistic regression as a starting point to design a neural network which would be at least as good as the logistic model is presented in Ciampi and Zhang [44] using ten medical data sets.

The extension to the polychotomous case is obtained considering $K > 1$ outputs. The neural network of Figure 5.2 is characterized by weights $\beta_{ij}(j = 1, \ldots, K)$ to output y_j, obtained by

$$p_j = \Lambda\left(\beta_{oj} + \sum_{i=1}^{k} \beta_{ij}x_i\right). \tag{5.9}$$

In statistical terms, this network is equivalent to the multinomial or polychotomous logistic model defined as

$$p_j = P(y_j = 1/\underset{\sim}{x}) = \frac{\Lambda(x, \beta_j)}{\sum_{j=1}^{k} \Lambda(x, \beta_j)}. \tag{5.10}$$

The weights can be interpreted as regression coefficients and maximum likelihood obtained by back propagation maximum likelihood.

There is a huge literature, especially in medical sciences, with comparisons among these two models (LR and ANN). Here, we will only give some useful references mainly from the statistical literature. This section ends with two applications showing some numerical aspects of the models equivalence as given in Schumacher et al. [189].

Applications and comparisons of interest are: personal credit risk scores [9], country international credit scores [48], polychotomous logistic for discrete choice [32], insolvency risk scores [73, 124, 28], and some medical applications [83, 157, 152, 22].

The first illustration is from Finney [70] and the data consists of $N = 39$ binary responses denoting the presence $(y = 1)$ or absence $(y = 0)$ of vasoconstriction in the finger's skin after the inhale of an air volume at a mean rate of inhale R.

Table 5.2 presents the results of the usual analysis from two logistic regression models. The first, considers only one covariate $x_1 = \log R$, that is

$$P(y = 1/x) = \Lambda(\beta_0 + \beta_1 x_1). \tag{5.11}$$

The second model considers the use of a second covariate, that is

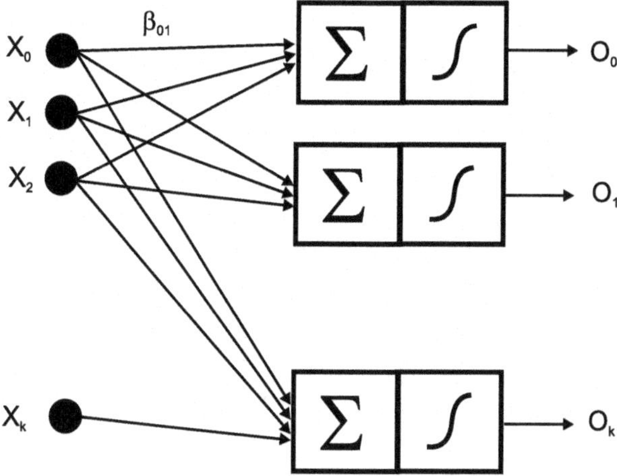

Figure 5.2
Polychotomous logistic network.

$$P(y|x) = \Lambda(\beta_0 + \beta_1 + \beta_2 x_2), \tag{5.12}$$

for $x_1 = \log$ and $x_2 = \log V$, respectively.

In the first case, the results show an effect marginally significant of the logarithm of rate of inhale in the probability to the presence of vasoconstriction in the finger's skin.

An ANN equivalent to this model would be a feedforward network with two input neurons ($x_0 = 1, x_1 = \log R$). Using a rate of learning $\eta = 0.001$ to the last interactions ML-BP, we have $w_0 = -0.469$ and $w_1 = 1.335$ with a Kullback-Leiber measure of $E^* = 24.554$.

For the case of the model with two covariates, the result shows a significant influence of the two covariates in the probability of vasoconstriction in the finger's skin. One perceptron with three input neurons ($x0 = 1, x1 = \log R, x2 = \log V$) should give similar results if ML-BP was used. This was almost obtained ($w_0 = 2.938, w_1 = 4.650, w_2 = 4.240$) although the algorithm backpropagation demands several changes in learning rate from an initial value $\eta = 0.2$ to $\eta = 0.000001$ involving approximately $20,000$ interactions to obtain a value of the distance Kullback-Leibler of $E^* = 14.73$ comparable to that of minus the log-likelihood ($L = -14.632$).

The second example is a study to examine the role of noninvasive sonographic measurements for differentiation of benign and malignant breast tumors. Sonography measurements in $N = 458$ women and characteristics $(x_1, \ldots, x_p)$ were collected.

Cases were verified to be $325(y = 0)$ benign and 133 as $(y = 1)$ malignant tumors. A preliminary analysis with a logistic model indicates three covariates as

Table 5.2
Results of a logistic regression analysis for the vasoconstriction data.

Variable	Coefficient of Regression	Standard Error	P-value	Neural Network
Intercept	-0.470	0.439	-	-0.469
Log R	1.336	0.665	0.045	1.335
		$(L = 24.543)$		$(E = 24.554)$
Intercept	-2.924	1.288	-	-2.938
Log R	-4.631	1.789	0.010	-4.651
Log V	5.221	1.858	0.005	5.240
		$(L = 14.632)$		$(E = 14.73)$

Table 5.3
Results of a logistic regression analysis of breast tumor data.

		Logistic		
Variable	Coefficients	Standard Error	P Value	Neural Network Weights
Intercept	-8.178	0.924		$w_0 = -8.108$
Age	0.070	0.017	0.0001	$w_1 = 0.069$
log AT+1	5.187	0.575	0.0001	$w_2 = 5.162$
log AC+1	-1.074	0.437	0.0014	$w_3 = -1.081$
		L=79.99		E=80.003

significant, this indication plus problems of collinearity among the six variables suggests the use of age, number of arteries in the tumor (AT), number of arteries in the contralateral breast (AC). The results are shown in Table 5.3.

A comparison with ANN is shown in the Table 5.4.

5.1.2 Regression Networks

The regression network is a GLIM model with identity link function. The introduction of hidden layer has no effect in the architecture of the model because of the linear properties and the architecture remains as in Figure 5.1.

Some authors have compared the prediction performance of econometric and regression models with feedforward networks usually with one hidden layer. In fact, they compared linear models with nonlinear models, in this case, a particular projection pursuit model with the sigmoid as the smooth function (see Section 5.2.5). An extension for system of equations can be obtained for systems of k equations by considering k outputs in the neural network as in Figure 5.2. Again, the outputs activation function is the identity function.

In what follows, we describe some comparisons of regression models versus neural networks.

Gorr et al. [80] compare linear regression, stepwise polynomial regression and

Table 5.4
Results of classification rules based on logistic regression, CART and feedforward neural networks with j hidden units (NN(J)) for the breast tumor data.

Method	Sensitivity (%)	Specificity (%)	Percentage of correct classification
Logistic regression	95.5	90.2	91.7
CART	97.9	89.5	91.9
NN(1)	95.5	92.0	93.0
NN(2)	97.0	92.0	93.4
NN(3)	96.2	92.3	93.4
NN(4)	94.7	93.5	93.9
NN(6)	97.7	94.8	95.6
NN(8)	97.7	95.4	96.1
NN(10)	98.5	98.5	98.5
NN(15)	99.2	98.2	98.5
NN(20)	99.2	99.1	99.1
NN(40)	99.2	99.7	99.6

a feedforward neural network with three units in the hidden layer with an index used by an admissions committee to predict students' grade point averages (GPAs) as qualifications for professional school admission. The dependent variable to be predicted for admissions decision is the total GPA of all courses taken in the last two years. The admission decision is based on the linear decision rule.

$$LDR = math + chem + eng + zool + 10Tot + 2Res + 4PTran, \qquad (5.13)$$

which are the grades in mathematics, chemistry, English, zoology; and total GPA. The other variables are binary and indicate residence, transfer of partial credits in prerequisites, and transfer of all prerequisite credits.

Although none of the empirical methods was statistically significant better than the admission committee index, the analysis of the weights and output of the hidden units identifies three different patterns in the data. The results showed very interesting comparisons by using quartiles of the predictions.

Church and Curram [41] compared forecasts of personal expenditures obtained by neural network and econometric models (London Business School, National Institute of Economic and Social Research, Bank of England and a Goldman Sachs equation). None of the models could explain the decreases in expense growth at the end of the 1980s and the beginning of the 1990s. A summary of this application of neural networks is the following:

- Neural networks use exactly the same covariates and observations used in each econometric model. These networks produced similar results to the econometric models.
- The neural networks used 10 neurons and a hidden layer, in all models.
- The dependent variable and the covariates were re-escalated to fall in the interval 0.2 to 0.8.
- A final exercise used a neural network with inputs of all variables from all models. This network was able to explain the fall in growth, but it is a network with too many parameters.
- Besides the forecast comparison, a sensitivity analysis was done on each variable, so that the network was tested with the mean value of the data and each variable was varied to verify the grade in which each variation affects the forecast of the dependent variable.

Another econometric model comparison using neural network and linear regression was given by Qi and Maddala [171]. They relate economic factors with the stock market and in their context conclude that a neural network model can improve the linear regression model in terms of predictability, but not in terms of profitability. For other econometric applications see Silva et al. [192] and Tkacz [209].

In medicine, Laper et al. [115] compare a neural network with linear regression in predicting birthweight from nine perinatal variables, which are thought to be related. Results show that seven of the nine variables, i.e., gestational age, mother's body-mass index (BMI), sex of the baby, mother's height, smoking, parity, and gravidity, are related to birthweight. They found that neural network performed slightly better than linear regression and found no significant relation between birthweight and each of the two variables: maternal age and social class.

An application of neural networks in fitting the response surface is given by Balkin and Lim [15]. Using simulated results for an inverse polynomial they showed that a neural network can be a useful model for response surface methodology.

5.2 Nonparametric Regression and Classification Networks

Here, we relate some nonparametric statistical methods with the models found in the neural network literature. An introductory view can be seen in Picton [166].

5.2.1 Probabilistic Neural Networks (PNNs)

Suppose we have C classes and want to classify a vector $\underset{\sim}{x}$ to one of this classes. The Bayes classifier is based on $p(k/\boldsymbol{x})\alpha\pi_k p_k(\boldsymbol{x})$, where π_k is the prior probability of an observation be from class $k, p_k(\boldsymbol{x})$. It is the probability of $\boldsymbol{x}$ being observed among those of class $C = k$.

In general, the probability π_k and the density $p(\cdot)$ have to be estimated from data. A powerful nonparametric technique to estimate these functions is based on kernel methods and proposed by Parzen (Ripley [177, p. 184]).

A kernel K is a bounded function with an integral one. A suitable example is the multivariate normal density function. We use $K(\boldsymbol{x}-\boldsymbol{y})$ as a measure of the proximity of $\boldsymbol{x}$ on $\boldsymbol{y}$. The empirical distribution of $\boldsymbol{x}$ within a group k gives mass $\frac{1}{m_k}$ to each of the samples. A local estimate of the density $p_k(\boldsymbol{x})$ can be found summing all these contributions with weight $K(\boldsymbol{x} - \boldsymbol{x}_2)$ that is Ripley [177, p. 182].

$$\hat{p}_j(\boldsymbol{x}) = \frac{1}{n_j}\sum_{i=1}^{n_j} K(\boldsymbol{x} - \boldsymbol{x}_i). \tag{5.14}$$

This is an average of the kernel functions centered on each example from the class. Then, we have from Bayes theorem that

$$\hat{p}(k/\boldsymbol{x}) = \frac{\pi_k \hat{p}_k(\boldsymbol{x})}{\sum_{j=1}^{C} \pi_j \hat{p}_j(\boldsymbol{x})} = \frac{\frac{\pi_k}{n_k}\sum_{[i]=k} K(x - x_k)}{\sum_i \frac{\pi_{[i]}}{n_{[i]}} K(x - x_i)}. \tag{5.15}$$

When the prior probability are estimated by $n_{k/n}$, (5.15) simplifies to

$$p(k/\boldsymbol{x}) = \frac{\sum_{[i]=k} K(\boldsymbol{x} - \boldsymbol{x}_i)}{\sum_i K(\boldsymbol{x} - \boldsymbol{x}_i)}. \tag{5.16}$$

This estimate is known as Parzen estimate and probabilistic neural network Patterson [161, p. 352].

In radial basis function networks with a Gaussian basis, it is assumed that the normal distribution is a good approximation to the cluster distribution. The kernel nonparametric method of Parzen approximates the density function of a particular class in a pattern space by a sum of kernel functions, and one possible kernel is the Gaussian function. As we have seen, the probability density function for a class is approximated by the following equation

$$p_k(\boldsymbol{x}) = \left(\frac{1}{2\pi^{n/2}\sigma^2}\right)\frac{1}{n_K}\sum_{j=1}^{n_k} e^{-(x-x_{kj})^2/2\sigma^2}, \tag{5.17}$$

where σ^2 has to be decided. Choosing a large value results in overgeneralization, choosing a value too small results in overfitting. The value of σ should be dependent on the number of patterns in a class. One function used is

$$\sigma = an_k^{-b} \; a > 0, 0 < b < 1. \tag{5.18}$$

A three-layer network like a radial basis function network (RBFN) network is used with each unit in the hidden layer centered on an individual item. Each unit in the output layer has weights of 1, and a linear output function, this layer adds all the outputs from the hidden layer that correspond to data from the same class together. This output represents the probability that the input data belong to the class represented by that unit. The final decision as to what class the data belongs to is simply the unit with the output layer with the largest value. If values are normalized they lie between 0 and 1. In some networks an additional layer is included that makes this a winner-takes-all decision; each unit has binary output.

Figure 5.3 shows a network that finds three classes of data. There are two input variables, and for the three classes of data there are four samples for class A, three samples for class B and two samples for class C; hence the number of neurons in the hidden layer.

The advantage of the PNN is that there is no training. The values for the weights in the hidden units (i.e., the centers of the Gaussian functions) are just the values themselves. If the amount of data is large, some clustering can be done to reduce the number of units needed in the hidden k layer.

An application of PNN to classification of electrocardiograms with a 46-dimensional vector input pattern, using data of 249 patients for training and 63 patients for testing is mentioned in Specht [197]. A comparison of PNN with other networks can be seen in Kim and Chun [111].

5.2.2 General Regression Neural Networks (GRNNs)

The general regression neural network was named by Specht [196] and it was already known in the nonparametric statistical literature as Nadaraya-Watson estimator.

The idea of GRNN is to approximate a function given a set of n outputs $y_1, \ldots, y_n$ and n vector of inputs $\boldsymbol{x}_1, \ldots, \boldsymbol{x}_n$ using the following equation from probability theory for the regression of y given $\boldsymbol{x}$

$$\bar{y}(\bar{\boldsymbol{x}}) = E(y/\boldsymbol{x}) = \frac{\int y f(\boldsymbol{x}, y) dy}{\int f(\boldsymbol{x}, y) dy}, \tag{5.19}$$

where $f(\boldsymbol{x}, y)$ is the joint probability density function of $(\boldsymbol{x}, y)$.

As in PNN, we approximate the conditional density function by the sum of Gaussian function. The regression of y on $\boldsymbol{x}$ is then given by

$$\hat{y}(\hat{\boldsymbol{x}}) = \frac{\sum_{j=1}^{n} y_j e^{-d_j^2/2\sigma^2}}{\sum_{j=1}^{n} e^{-d_j^2/2\sigma^2}} \tag{5.20}$$

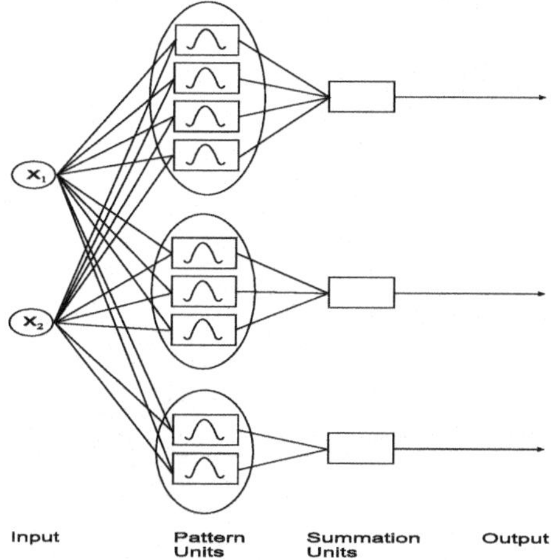

Figure 5.3
Probabilistic neural network (PNN).

as a particular form of the Nadaraya-Watson estimator:

$$\hat{y}(x) = \frac{\sum y_i K(\boldsymbol{x} - \boldsymbol{x}_i)}{\sum K(\boldsymbol{x} - \boldsymbol{x}_i)}. \tag{5.21}$$

The value of d_j is the distance between the current input and the j-th input in the sample of training set. The radial σ is again given by

$$\sigma = an^{-b/n}. \tag{5.22}$$

The architecture of the GRNN is similar to PNN, except that the weights in the output layer are not set to 1, but instead they are set to the corresponding values of the output y in the training set. In addition, the sum of the outputs from the Gaussian layer has to be calculated so that the final output can be divided by the sum. The architecture is shown in Figure 5.4 for a single output of two variables input with a set of 10 data points.

5.2.3 Generalized Additive Model Networks

The generalized additive model (GAM) relating an output y to a vector input $\boldsymbol{x} = (x_1, \ldots, x_I)$ is given by

$$E(Y) = \mu, \tag{5.23}$$

$$h(\mu) = \eta = \sum_{i=0}^{I} f_i(x_i). \tag{5.24}$$

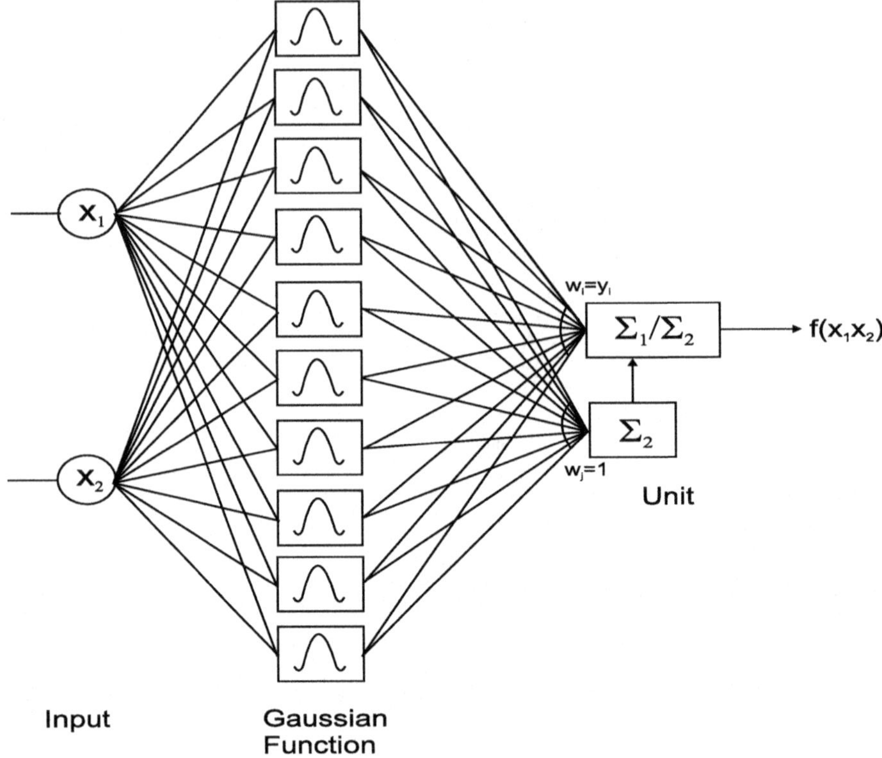

Figure 5.4
General regression neural network architecture.

This is a generalization of the GLIM model when the parameters β_i are replaced by functions f_i. Again $h(\cdot)$ is the link function. Nonparametric estimation is used to obtain the unknown functions f_j. They are obtained iteratively by first fitting f_1, then fitting f_2 to the residual $y_i - f_1(x_{1i})$, and so on. Ciampi and Lechevalier [43] used the model

$$logit(p) = f_1(x_1) + \ldots + f_I(x_I) \tag{5.25}$$

as a 2 class classifier. They estimated the functions f using B-splines.

A neural network corresponding to (5.25) with two continuous inputs is shown in Figure 5.5. The first hidden layer consists of two blocks, each corresponding to the transformation of the input with B-splines. A second hidden layer computes the f's functions; all units have as activation the identity function. The output layer has a logistic activation function and output p.

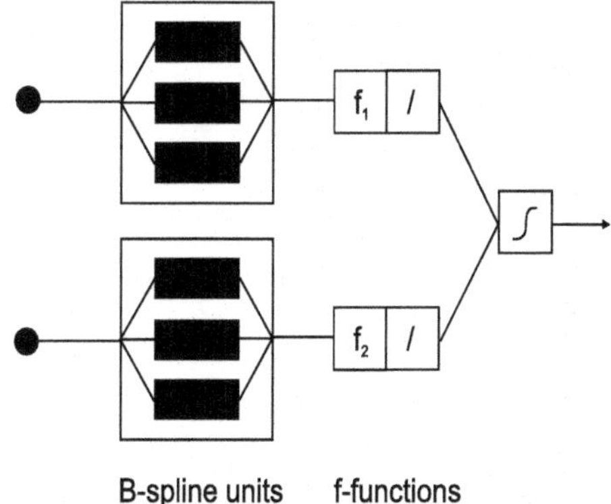

Figure 5.5
GAM neural network.

5.2.4 Regression and Classification Tree Networks

Regression tree is a binning and averaging procedure. The procedure is presented in Fox [75] and Fox [74] for a sample regression of y on x, to average the values of y for corresponding values of x for certain conveniently chosen regions. Some examples are shown in Figure 5.6.

Closely related to regression trees are classification trees, where the response variable is categorical rather than quantitative.

The regression and classification tree models can be written respectively as:

$$Y = \alpha_1 I_1(\boldsymbol{x}) + \ldots + I_L(\boldsymbol{x}), \tag{5.26}$$

$$logit(p) = \gamma_1 I_1(\boldsymbol{x}) + \ldots + I_L(\boldsymbol{x}), \tag{5.27}$$

where the I's are characteristic functions of L-subsets of the predictor space, which form a partition.

The following classification tree is given in Ciampi and Lechevalier [43].

The leaves of the tree represent the sets of the partition. Each leaf is reached through a series of binary questions involving the classifiers (input); these are determined from the data at each mode.

The trees can be represented as neural networks. Figure 5.8 shows such representation for the tree of Figure 5.7.

The hidden layers have activation function taking value -1 for negative input and 1 for positive input. The weights linking the second hidden layer to the output are determined by the data and are given by the logit of the class probability corresponding to the leaves of the tree.

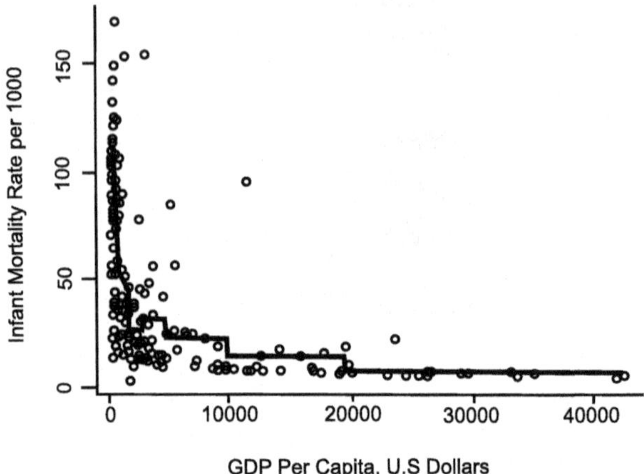

(a) The binning estimator applied to the relationship between infant mortality per 1000 and GDP per capita, in US dollars. Ten bins are employed.

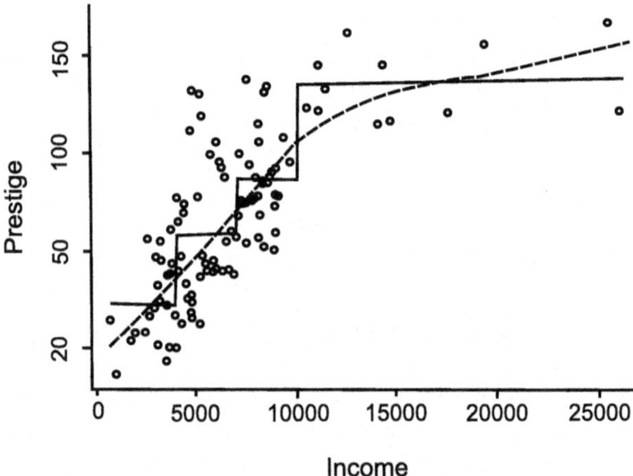

(b) The solid line gives the estimated regression function relating occupational prestige to income implied by the tree; the broken line is for a local linear regression.

Figure 5.6
Regression tree $\hat{y}$.

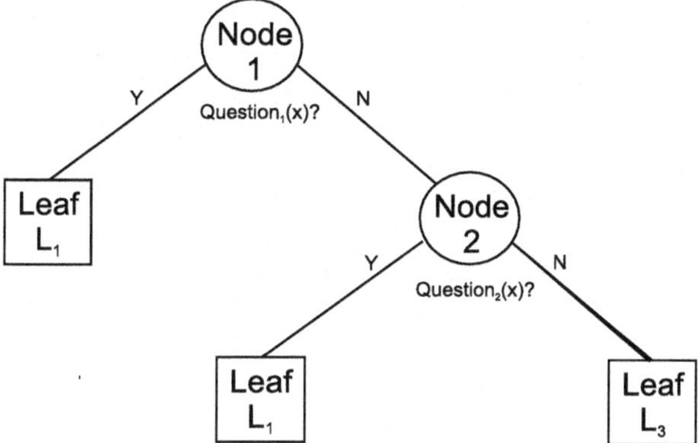

Figure 5.7
Decision tree.

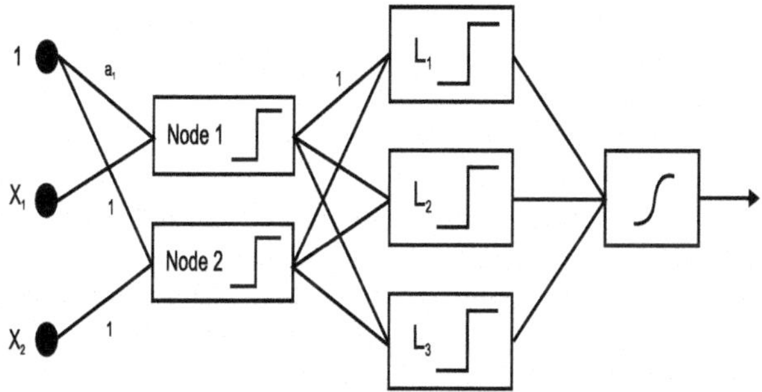

Figure 5.8
Classification tree network.

5.2.5 Projection Pursuit and Feedforward Networks

Generalized additive models fit

$$y_i = \alpha + f_1(x_1) + \ldots + f_I(x_I). \tag{5.28}$$

Projection pursuit regression (PPR), models fit

$$y_i = \alpha + f_1(z_{i1}) + \ldots + f_I(z_{iI}), \tag{5.29}$$

where the z_1's are linear combinations of the x's

$$z_{ij} = \alpha_{j1}x_{i1} + \alpha_{j2}x_{i2} + \ldots + \alpha_{\alpha p}x_{ip} = \boldsymbol{\alpha}_j'\boldsymbol{x}_i. \tag{5.30}$$

The projection-pursuit regression model can therefore capture some interactions among the x's. The fitting of this model is obtained by least squares fitting of $\hat{\alpha}_1$,

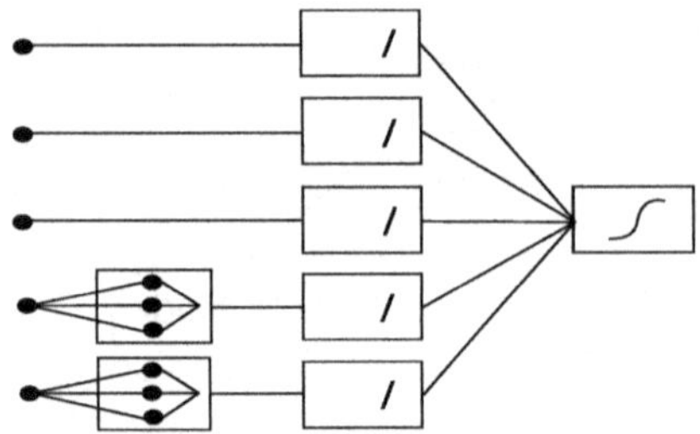

Figure 5.9
GAM network.

then as in GAM fitting $\hat{f}_1(\hat{\alpha}_1'\boldsymbol{x})$ from the residual R_1 of this model, estimate $\hat{\alpha}_2$ and then $\hat{f}_2(\hat{\boldsymbol{\alpha}}_2\boldsymbol{x})$ and so on.

The model can be written as

$$y_k = f(\boldsymbol{x}) = \alpha_k + \sum_{j=1}^{I} f_{ik}(\alpha_j + \boldsymbol{\alpha}_j'\boldsymbol{x}) \tag{5.31}$$

for multiple output $\boldsymbol{y}$.

Equation (5.31) is for a feedforward network, if, for instance, the functions f_j are sigmoid function and we consider the identity output.

Clearly feedforward neural network is a special case of PPR taking the smooth functions f_j to be logistic functions. Conversely, in PPR, we can approximate each smooth function f_i by a sum of logistic functions.

Another heuristic reason why feedforward network might work well with a few hidden neurons, for approximation of functions, is that the first stage allows a projection of the $\boldsymbol{x}$'s onto a space of much lower dimension for the z's.

5.2.6 Example

Ciampi and Lechevalier [43] analyzed a real data set of 189 women who had babies. The output variable was 1 if the mothers delivered babies of normal weight and 0 for mothers who delivered babies of abnormally low weight (< 2500 g). Several variables were observed in the course of pregnancy: two continuous variables (age in years and mother's weight in last menstrual period before pregnancy) and six binary variables.

They apply a classification tree network, a GAM network, and a combination of the two. Figure 5.8 gives their classification network. Their GAM and combined network are shown in Figures 5.9 and 5.10. The misclassification numbers from each network were 50 for GAM; 48 for classification tree, and 43 for network of networks.

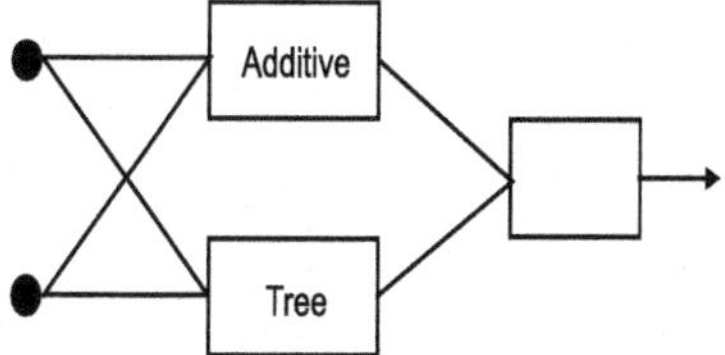

Figure 5.10
Network of networks.

5.3 Regression Neural Network Models with Python

In this section, we have a model for classification and cross validation with a probabilistic neural network, based on concepts similar to those of the multilayer feed-forward networks presented in Section 3.7.2. Similarly, we create a dataset with 20 points, where 10 represent one type and the others 10 represent another type. We use 5 points of each type for training the PNN and the other 5 points for testing the network. Thus, the PNN tries to predict whether the point is of type 1 or 0. In our example, Program 3.4 predicts 8 corrects and 3 wrongs. The vectors of prediction, types, and errors are as follows.

```
Prediction:        [0 1 0 0 1 0 0 1 1 0]
Types:             [0 1 0 0 1 1 0 0 1 1]
Errors:            [0 0 0 0 0 1 0 1 0 1]
```

Since we set `shuffle=False` at the end of line 12 and use the same parameters, the dataset generated is equal to the dataset in Program 3.4. To generate different datasets and outputs, set `shuffle=True`. Compare Program 5.1 with Program 3.4 and the Figure 5.11 with the Figure 3.21.

Program 5.1
Classification and cross validation with a PNN.

```
1 import numpy as np
2 from matplotlib import pyplot as plt
3 from matplotlib.colors import ListedColormap as cm
4 from sklearn import datasets, model_selection
5 from neupy.algorithms import PNN
6
7 sample, types = datasets.make_moons( noise=.5,
8             n_samples=20, random_state=0)
```

```
9
10 sample_train, sample_test, types_train, types_test \
11          =model_selection.train_test_split(
12          sample, types, test_size=10, shuffle=False)
13
14 pnn = PNN(std=0.1)
15 pnn.train(sample_train, types_train)
16 Z = pnn.predict(sample_test)
17
18 plt.scatter(sample_train[:, 0], sample_train[:, 1],
19              marker='.', c=types_train,
20              cmap=cm(['#050505', '#AAAAAA']))
21
22 plt.scatter(sample_test[:, 0], sample_test[:, 1],
23              marker='x', c=types_test,
24              cmap=cm(['#050505', '#AAAAAA']))
25
26 print("Prediction:\t",Z)
27 print("Types:\t\t",types_test)
28 S=np.logical_xor(Z,types_test, casting='same_kind')
29 S=S*1
30 print("Errors:\t\t",S)
31
32 plt.scatter(sample_test[:, 0], sample_test[:, 1],
33              marker='o', s=S*200,
34              color='#050505', facecolors='none')
35
36 plt.show()
```

Line 32 in Program 5.1 highlights the errors with circles in the plot presented in Figure 5.11.

Program 5.2 performs the same function as Program 5.1, but instead of using a PNN, it uses a GRNN. In fact, these networks are best suited for regression problems.

Program 5.2

Classification and cross validation with a GRNN.

```
import numpy as np
from matplotlib import pyplot as plt
from matplotlib.colors import ListedColormap as cm
from sklearn import datasets, model_selection
from neupy.algorithms import GRNN

sample, types = datasets.make_moons( noise=.5,
          n_samples=20, random_state=0)
```

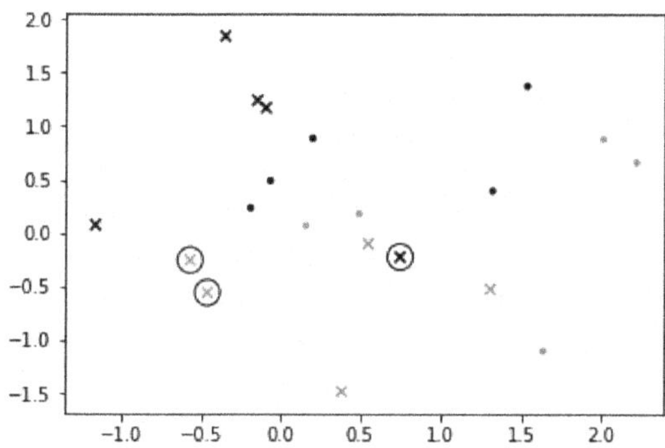

Figure 5.11
Classification and cross validation with a PNN.

```
sample_train, sample_test, types_train, types_test \
        =model_selection.train_test_split(
        sample, types, test_size=10, shuffle=False)

grnn = GRNN(std=0.1)
grnn.train(sample_train, types_train)
Z = grnn.predict(sample_test)

plt.scatter(sample_train[:, 0], sample_train[:, 1],
            marker='.', c=types_train,
            cmap=cm(['#050505', '#AAAAAA']))

plt.scatter(sample_test[:, 0], sample_test[:, 1],
            marker='x', c=types_test,
            cmap=cm(['#050505', '#AAAAAA']))

Z=(np.round(Z[:,0])>0)*1
print("Prediction:\t",Z)
print("Types:\t\t",types_test)
S=np.logical_xor(Z,types_test, casting='same_kind')
S=S*1
print("Errors:\t\t",S)

plt.scatter(sample_test[:, 0], sample_test[:, 1],
            marker='o', s=S*200,
            color='#050505', facecolors='none')

plt.show()
```

Program 5.3
Approximation of $\sin(x) + x$ with a GRNN.

```
import numpy as np
import matplotlib.pyplot as plt
from neupy.algorithms import GRNN

x = np.arange(0, 10, .005).reshape(-1, 1)
y = (np.sin(x)+x).ravel()

grnn = GRNN(std=0.1)
grnn.train(x, y)

test_x = np.arange(0, 10, 0.5).reshape(-1, 1)
test_y = grnn.predict(test_x)

plt.scatter(x, y, c='gray', marker=".")
plt.scatter(test_x,test_y, s=25, c='black', marker="D")

plt.xlim(0, 10)
plt.ylim(0, 10)

plt.show()
```

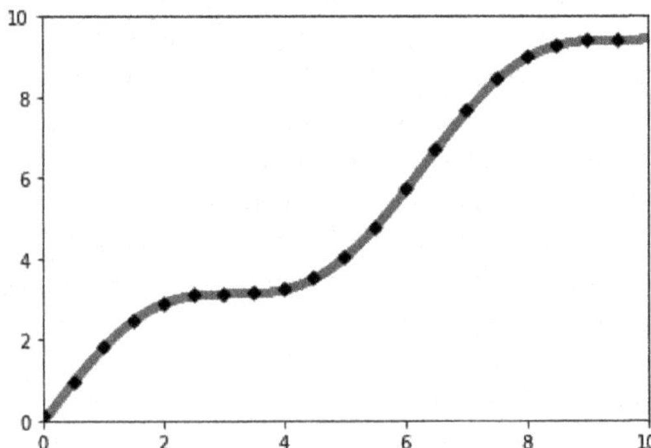

Figure 5.12
Approximation of $\sin(x) + x$ with a GRNN.

The output from Program 5.3 in Figure 5.12 is much better than the outputs from Program 4.4 in Figure 4.40 and the Program 3.2 in Figure 3.19.

6

Survival Analysis and Other Networks

Further statistical methods of data analysis are studied using neural networks: survival analysis networks, time series forecasting networks, and quality control charts networks. Statistical inferences are briefly mentioned, namely, point estimation methods, interval estimation, statistical tests, and Bayesian methods.

6.1 Survival Analysis Networks

If T is a non-negative random variable representing the time to failure or death of an individual, we may specify the distribution of T by any one of the probability density function $f(t)$, the cumulative distribution function $F(t)$, the survivor function $S(t)$, the hazard function $f(t)$, or the cumulative hazard function $H(t)$. These are related by

$$
\begin{aligned}
F(t) &= \int_0^t f(u)du \\
f(t) &= F'(t) = \frac{d}{dt}F(t) \\
S(t) &= 1 - F(t) \\
h(t) &= \frac{f(t)}{S(t)} = -\frac{d}{dt}[\ln S(t)] \\
H(t) &= \int_0^t h(u)du \\
h(t) &= H'(t) \\
S(t) &= \exp[-H(t)].
\end{aligned}
\tag{6.1}
$$

A distinct feature of survival data is the occurrence of incomplete observations. This feature is known as *censoring* and it may arise from time limits or other restrictions, depending on the study.

There are different types of censoring.

- Right censoring occurs if the events are not observed before the prespecified study-term or some competitive event (e.g., death by other cause) that causes interruption of the follow-up on the individual experimental unit.

- Left censoring happens if the starting point is located before the time of the beginning of the observation for the experimental unit (e.g., time of infection by HIV virus in a study of survival of AIDS patients).

- Interval censoring occurs if the exact time of an event is unknown, but we know that it falls within an interval I_i (e.g., when observations are grouped).

The aim is to estimate the previous functions from the observed survival and censoring times. This can be done either by assuming some parametric distribution for T or by using non-parametric methods. Maximum likelihood techniques can fit parametric models of survival distributions. The usual non-parametric estimator for the survival function is the Kaplan-Meier estimate. When two or more group of patients are to be compared the log-rank or the Mantel-Hanszel tests are used.

General classes of densities and the non-parametric procedures with estimation procedures are described in Kalbfleish and Prentice [106].

Usually, we have covariates related to the survival time T. The relation can be linear ($\boldsymbol{\beta}'\boldsymbol{x}$) or nonlinear ($g(\gamma;\boldsymbol{x})$). A general class of models relating survival time and covariates is studied in Louzada-Neto [131] and Louzada-Neto [132]. Here, we describe the three most common particular cases of the Louzada-Neto model.

The first class of models is the *accelerated failure time* (AFT) models

$$\log T = -\boldsymbol{\beta}'\boldsymbol{x} + W, \tag{6.2}$$

where W is a random variable. Then exponentiation gives

$$T = \exp\left(-\boldsymbol{\beta}'\boldsymbol{x}\right) e^{W} \text{ or } T' = e^{W} = T\exp\left(\boldsymbol{\beta}'\boldsymbol{x}\right), \tag{6.3}$$

where T' has hazard function h_0 that does not depend on β. If $h_j(t)$ is the hazard function for the j'th patient it follows that

$$h_j(t) = h_0(t\exp\boldsymbol{\beta}'\boldsymbol{x})\exp\boldsymbol{\beta}'\boldsymbol{x}. \tag{6.4}$$

The second class is the *proportional odds* (PO) where the regression is on the log-odds of survival, correspondence to a linear logistic model with either death or no death by a fixed time t_0' as a binary response.

$$\log\frac{S_j(t)}{1-S_j(t)} = \boldsymbol{\beta}'\boldsymbol{x} + \log\frac{S_0(t)}{1-S_0(t)} \tag{6.5}$$

or

$$\frac{S_j(t)}{1-S_j(t)} = \frac{S_0(t)}{1-S_0(t)}\exp\boldsymbol{\beta}'\boldsymbol{x}. \tag{6.6}$$

The third class is the *proportional hazard* or *Cox regression model* (PH) given by

$$\log h_t(t) = \boldsymbol{\beta}'\boldsymbol{x} + \log h_0(t), \tag{6.7}$$

$$h_j(t) = h_0(t)\exp\boldsymbol{\beta}'\boldsymbol{x}. \tag{6.8}$$

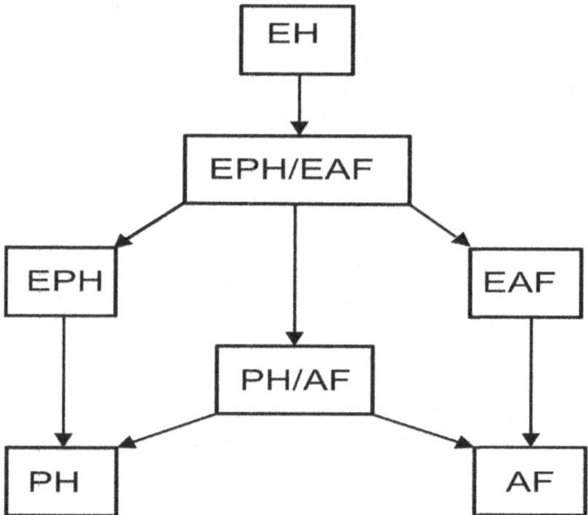

Figure 6.1
Classes of regression models for survival data [131]. AF = accelerated failure. PH/AF = mixed model. EPH = extended PH. EAF = extended AF. EPH/EAF = extended mix. EH = extended hazard.

Ciampi and Etezadi-Amoli [42] extend models (6.2) and (6.7) as a mixed model. They extend the two models and also include three more models to form a more comprehensive model [133]. See Figure 6.1.

Ripley [179] investigated seven neural networks in modeling breast cancer prognosis; her models were based on alternative implementation of models (6.2) to (6.5) and allowed for censoring. Here, we outline the important results of the literature.

The accelerated failure time (AFT) model is implemented using the architecture of regression network with the censored times estimated using some missing value method as in Xiang et al. [226].

For the Cox proportional hazard model, Faraggi and Simon [66] substitute the linear function βx_j by the output $f(x_j, \theta)$ of the neural network, that is

$$L_c(\theta) = \prod_{i\in\ldots} \frac{\exp\left(\sum_{h=1}^{H} \alpha_h/[1+\exp(-w_h' x_i)]\right)}{\sum_{j\in R_i} \exp\left(\sum_{h=1}^{H} \alpha_h/[1+\exp(-w_h' x_i)]\right)} \tag{6.9}$$

and estimations are obtained by maximum likelihood through Newton-Raphson.

The corresponding network is shown in Figure 6.2.

As an example, Faraggi and Simon [66] consider the data related to 506 patients with prostatic cancer in stages 3 and 4. The covariates are: stage, age, weight, and treatment (0.2; 1 or 5 mg of diethylstilbestrol or placebo).

The results are given in the Tables 6.1, 6.2, and 6.3 below for the models:

(a) First-order PH model (4 parameters);

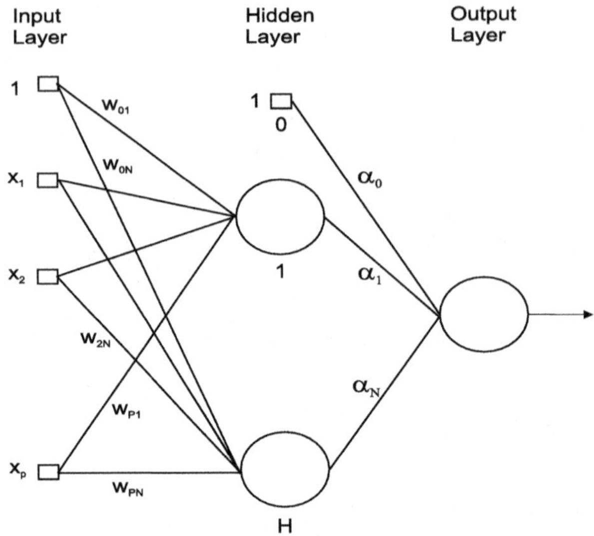

Figure 6.2
Neural network model for survival data (single hidden layer neural network).

(b) Second-order (interactions) PH model (10 parameters);

(c) Neural network model with two hidden nodes (12 parameters);

(d) Neural network model with three hidden nodes (18 parameters).

Table 6.1
Summary statistics for the factors included in the models.

	Complete data	Training set	Validation set
Sample size	475	238	237
Stage 3	47.5%	47.6%	47.4%
Stage 4	52.5%	52.4%	52.6%
Median age	73 years	73 years	73 years
Median weight	98-0	97-0	99-0
Treatment: Low	49.9%	48.3%	51.5%
Treatment: High	50.1%	51.7%	48.5%
Median survival	33 months	33 months	34 months
% censoring	28.8%	29.8%	27.7%

Implementation of the proportional odds and proportional hazards were implemented also by Liestol et al. [126] and Biganzoli et al. [18].

Table 6.2
Log-likelihood and c statistics for first-order, second-order, and neural network proportional hazard models.

Model	Number of parameter s	Training data Log lik	Training data c	Test data Log lik	Test data c
First order PH	4	-814.3	0.608	-831.0	0.607
Second-order PH	10	-805.6	0.648	-834.8	0.580
Neural network $H = 2$	12	-801.2	0.646	-834.5	0.600
Neural network $H = 3$	18	-794.9	0.661	-860.0	0.582

Table 6.3
Estimation of the main effects and higher order interactions using 2^4 factorial design contrasts and the predictions obtained from the different models.

Effects	PH 1st order	PH 2nd order	Neural network $H = 2$	Neural network $H = 3$
Stage	0.300	0.325	0.451	0.450
Rx^*	-0.130	-0.248	-0.198	-0.260
Age	0.323	0.315	0.219	0.278
Weight	-0.249	-0.238	-0.302	-0.581
Stage $\times$ Rx	0	-0.256	-0.404	-0.655
State $\times$ Age	0	-0.213	-0.330	-0.415
State $\times$ Wt^*	0	-0.069	-0.032	-0.109
Rx $\times$ Age	0	0.293	0.513	0.484
$Rx \times Wt$	0	-0.195	-0.025	0.051
Stage $\times$ Rx $\times$ Age	0	0	0.360	0.475
Stage $\times$ $Rx \times Wt$	0	0	0.026	0.345
Stage $\times$ Age $\times$ Wt	0	0	-0.024	0.271
Rx $\times$ Age $\times$ Wt	0	0	0.006	-0.363
State $\times$ Rx $\times$ Age $\times$ Wt	0	0	0.028	-0.128

* Rx = Treatment, Wt = Weight.

Liestol et al. [126] used a neural network for Cox's model with covariates in the form below.

Let T be a random survival time variable, and I_k the interval $t_{k-1} < t < t_k,\ k = 1, \ldots, K$ where $0 < t_0 < t_1 < \ldots < t_k < \infty$.

The conditional probabilities can specify the model

$$P(T \in I_k / T > t_{k-1}, x) = \frac{1}{1 + \exp\left(-\beta_{0k} - \sum_{i=1}^{1} \beta_{ik} x_i\right)} \tag{6.10}$$

for $K = 1, \ldots, K$.

The corresponding neural network is the multinomial logistic network with k outputs, see the Figure 6.3.

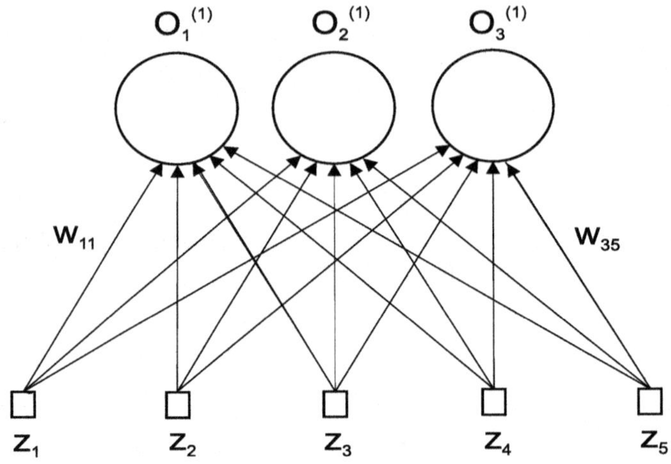

Figure 6.3
Log odds survival network.

The output 0_k in the kth output neuron corresponds to the conditional probability of dying in the interval I_k.

Data for the individual n consist of the regressor x^n and the vector $(y_1^n, \ldots, y^n)$ where y_k^n is the indicator of individual n dying in I_k and $k_n \leq K$ is the number of intervals where n is observed. Thus, $y_1^n, \ldots, y^n k_{n-1}$ are all 0 and $y^n k_n = 1$ of n dies in I_{kn} and

$$0_k = f(x, w) = \Lambda\left(\beta_{0k} + \sum_{i=1}^{1} \beta_{ik} x_i\right), \tag{6.11}$$

and the function to optimize

$$E^*(w) = \sum_{h=1}^{N} \sum_{k=1}^{K_n} -\log\left(1 - |y_k^n - f(x^n, w)|\right) \tag{6.12}$$

and $w = (\beta_{01}, \ldots, \beta_{0k}, \beta_{11}, \ldots, \beta_{Ik})$ and under the hypothesis of proportional rates make the restriction $\beta_{1j} = \beta_2 = \beta_{3j} = \beta_{4j} = \ldots = \beta_j$. Other implementations can be seen in Biganzoli et al. [18].

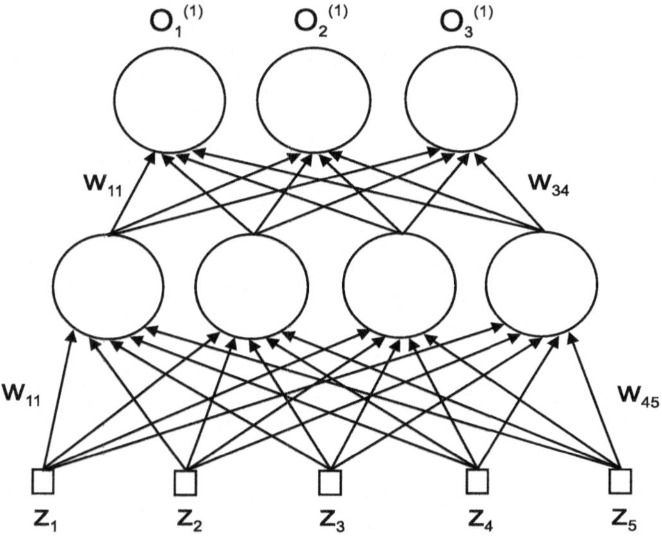

Figure 6.4
Nonlinear survival network. Two-layer feedforward neural networks showing input nodes and weights: input nodes (•), output (and hidden) nodes (O), covariates Z_j, connection weights w_{ij}, output values $0_1^{(2)}$.

An immediate generalization would be to substitute the linearity for nonlinearity of the regressors adding a hidden layer as in the Figure 6.4.

An example from Liestol et al. [126] uses the data from 205 patients with melanoma of which 53 died; 8 covariates were included.

Several networks were studied and the negative of the likelihood (interpreted as prediction error) is given in the Table 6.4 below.

Table 6.4
Cross validation of models for survival of patients with malignant melanoma.

	1	2	3	4	5	6	7
Prediction error	172.6	170.7	168.6	181.3	167.0	168.0	170.2
Change		-1.9	-4.0	8.7	-5.6	-4.6	-2.4

Column 1 = linear model. Column 2 = linear model with weight decay. Column 3 = linear model with penalty terms for non-proportional hazards. Column 4 = Non-linear model with proportional hazards. Column 5 = Non-linear model with penalty terms for non-proportional hazards. Column 6 = Non-linear model with proportional hazards in first and second intervals and in third and fourth intervals. Column 7 = Non-linear model with non-proportional hazards.

The main results for nonlinear models with two hidden nodes were:

- Proportional hazard models produced inferior predictions, decreasing the test log-likelihood of a two hidden node model by 8.7 (column 4) when using the standard weight decay, even more if no weight decay was used.
- A gain in the test log-likelihood was obtained by using moderately non-proportional models. Adding a penalty term to the likelihood of a non-proportional model or assuming proportionality over the two first- and last-time intervals improved the test log-likelihood by similar amounts (5.6 in the former case (column 5) and 4.6 in the latter (column 6)). Using no restrictions on the weights except weight decay gave slightly inferior results (column 7, improvement 2.4).

In summary, for this small data set the improvements that could be obtained compared to the simple linear models were moderate. Most of the gain could be obtained by adding suitable penalty terms to the likelihood of a linear but non-proportional model.

An example from Biganzoli et al. [18] is the application neural networks in the data sets of Efron's brain and neck cancer and Kalbfleish and Prentice lung cancer studies using the network architecture of Figure 6.5. The results of the survival curve fits follow in Figures 6.6 and 6.7.

A further reference is Bakker et al. [12] who used a neural-Bayesian approach to fit Cox survival model using Markov chain Monte Carlo (MCMC) and an exponential activation function. Other applications can be seen in Lapuerta et al. [116], Ohno-Machado et al. [155], Mariani et al. [139], Biganzoli et al. [19], Groves et al. [81] Faraggi et al. [67, 64], Ohno-Machado [154], Ripley and Ripley [178], and Pereira and R. [165].

6.2 Time Series Forecasting

A time series is a set of data observed sequentially in time; although time can be substituted by space, depth, etc. Neighboring observations are dependent, and the study of time series data consists in modeling this dependency.

A time series is represented by a set of observations $\{\boldsymbol{y}(t), \boldsymbol{t} \in T\}$, where T is a set of indices (time, space, etc.).

The nature of Y and T can be each either a discrete or continuous set. Further, $\boldsymbol{y}$ can be univariate or multivariate and T can have unidimensional or multidimensional elements. For example, $(Y, Z)'_{(t,s)}$ can represent the number of cases Y of influenza and Z of meningitis per week t and state s in 2002 in the USA.

There are two main goals in the study of time series: first, to understand the underlying mechanism that generates the time series, and second, to forecast future values of it. Problems of interest are:

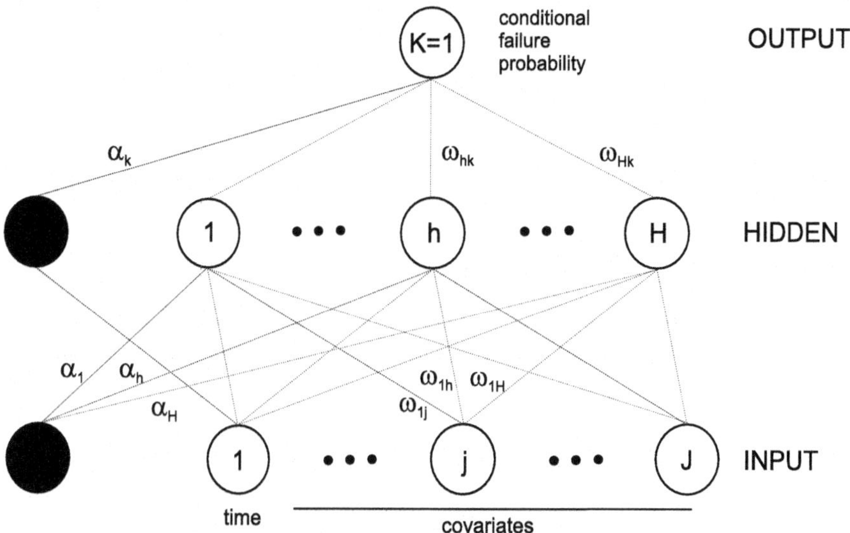

Figure 6.5
Partial logistic regression neural network (PLANN). The units (nodes) are represented by circles and the connections between units are represented by dashed lines. The input layer has J units for time a and covariates plus one unit (0). The hidden layer has H units plus the unit (0). A single output unit $(K = 1)$ computes conditional failure probability x_1, and x_2 are the weights for the connections of the unit with the hidden and output unit w_a, and the w's are the weights for the connections between input and hidden units and hidden and output unit, respectively.

- To describe of the behavior in the data.
- To find periodicities in the data.
- To forecast the series.
- To estimate the transfer function $v(t)$ that connects an input series X to the output series Y of the generating system. For example, in the linear case

$$Y(t) = \sum_{u=0}^{\infty} u(t)X(t-u). \tag{6.13}$$

- To forecast Y given v and X.
- To study the behavior of the system, by simulating X.
- To control Y by making changes in X.

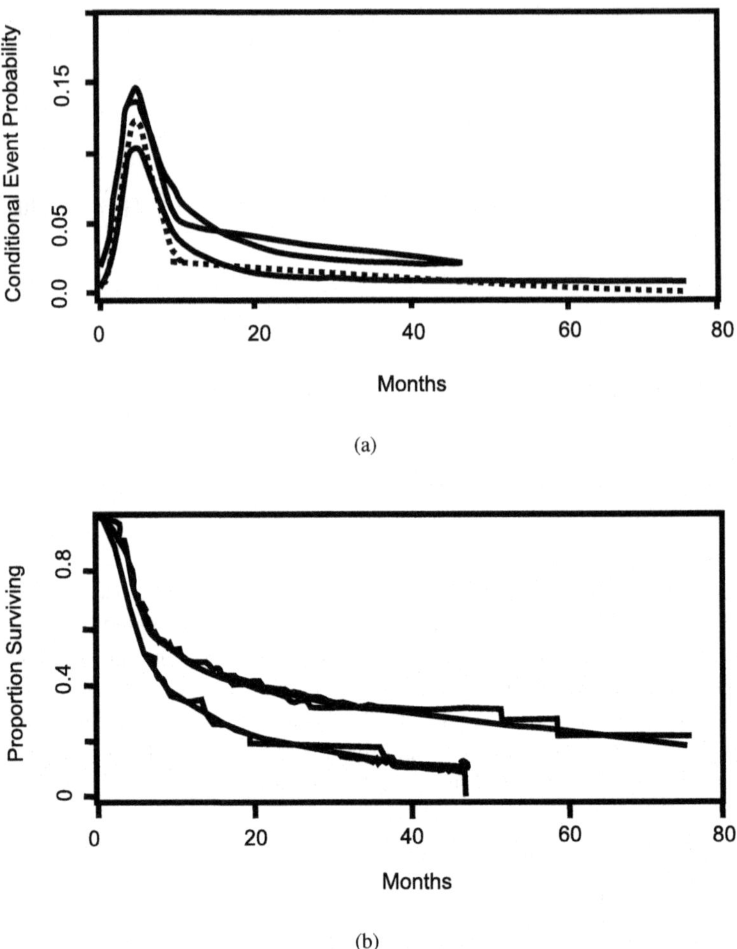

Figure 6.6
Best models of head and neck cancer trial. (a) Estimates of conditional failure probability obtained with the best PLANN configuration (solid line) and cubic linear spline proposed by Efron (dashed line). (b) Corresponding survival function and Kaplan-Meyer estimates.

There are basically two approaches for the analysis of time series. The first one analyzes the series in the time domain; that is, we are interested in the magnitude of the events that occur in time t. The main tool used is the autocorrelation function (and functions of it) and we use parametric models. In the second approach, the analysis occurs in the frequency domain; that is, we are interested in the frequency in

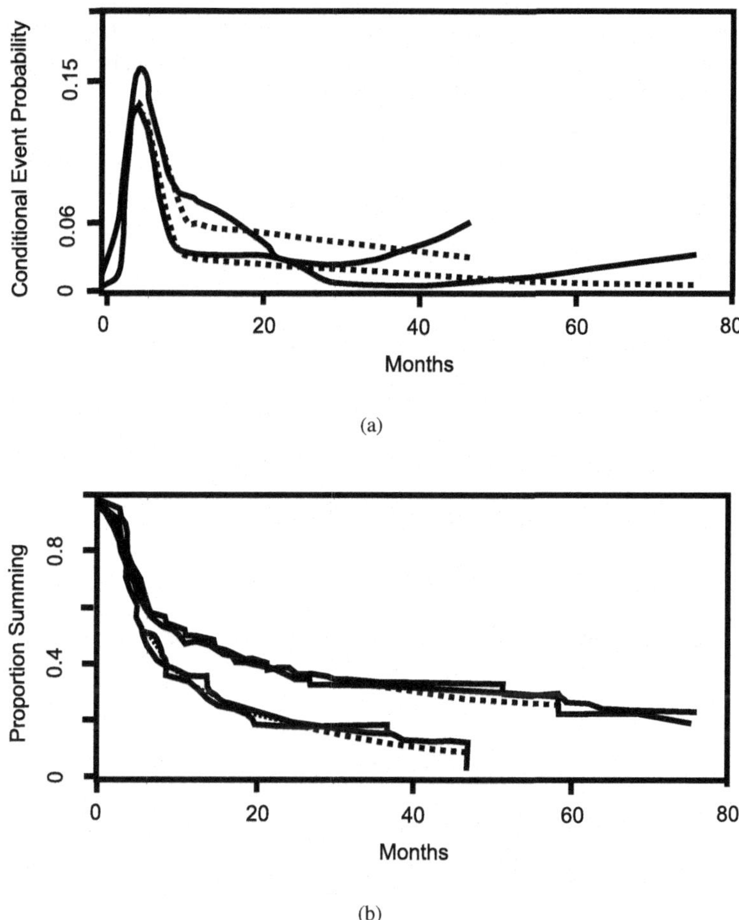

Figure 6.7
Suboptimal models of head and neck cancer trial. (a) Estimates of conditional failure probability obtained with the best PLANN configuration (solid line) and cubic linear spline proposed by Efron (dashed line). (b) Corresponding survival function and Kaplan-Meyer estimates.

which the events occur in a period (of time). The tool used is the spectrum (which is a transform of the correlation function, e.g., Walsh, Fourier, wavelet transforms) and we use non-parametric methods. The two approaches are not alternatives but complementary and are justified by representation theorems due to Wold [224] in time domain and Cramer [53] in the frequency domain. We can say from these theorems that the time domain analysis is used when the main interest is in the analysis of *non-deterministic* characteristic of this data and the frequency domain is used when the interest is in the *deterministic* characteristic of the data. Essentially Wold [224] results for the time domain state that any stationary process Y_t can be represented by

$$Y_t = D_t + Z_t, \quad (6.14)$$

where D_t and Z_t are uncorrelated, and D_t is deterministic, that is, it depends only on its past values $D_{t-1}, D_{t-2}, \ldots$ and Z_t is of the form

$$Z_t = \epsilon_t + \psi_1\epsilon_{t-1} + \psi_2\epsilon_{t-2} \quad (6.15)$$

or

$$Z_t = \pi_1 Z_{t-1} + \pi_2 Z_{t-2} + \ldots + \epsilon_t, \quad (6.16)$$

where ϵ_t is a white noise sequence (mean zero and autocorrelations zero) and ψ_1, π_1 are parameters. Equation (6.15) is an infinite moving-average process, $MA(\infty)$ and Equation (6.16) is an infinite autoregressive process $AR(\infty)$.

The frequency domain analysis is based on the pair of results

$$\rho(t) = \int_{-\infty}^{\infty} e^{iwt} dF(w), \quad (6.17)$$

where $\rho(t)$ is the autocorrelation of Y_t and acts as the characteristic function of the (spectral) density $F(w)$ of the process $Z(w)$ and

$$Y(t) = \int_{-\pi}^{\pi} e^{iwt} dZ(w). \quad (6.18)$$

Most of the results in time series analysis are based on these results from its extensions.

We can trace developments in time series chronologically over the 20th century, as shown in the Table 6.5.

Table 6.5
Historical development of the discipline.

Development	Discipline	Decades
Decomposition	Economics	1930s to 1950s
Exponential smoothing	Operations research	1960s
Box-Jenkins method	Engineering, statistics	1970s
State-space methods, Bayesian, Structural	Control engineering, statistics, and econometrics	1980s
Non-linear models	Statistics, control engineering	1990s

It should be pointed out that the importance of Box-Jenkins methodology is that they popularized and made accessible the difference equation representation of Wold. Further we can make an analogy between scalar and matrices and the difference equation models of Box-Jenkins (scalar) and state-space models (matrices). The recent interest in a nonlinear model for time series is that it makes neural networks attractive and we turn to this in the next section.

6.2.1 Forecasting with Neural Networks

The most commonly used architecture is a feedforward network with p lagged values of the time series as input and one output. As in the regression model of Section 5.1.2, if there is no hidden layer and the output activation is the identity function the network is equivalent to a linear $AR(p)$ model. If there are one or more hidden layers with a sigmoid activation function in the hidden neurons the neural network acts as nonlinear autoregression.

The architecture and implementation of most time series applications of neural networks follow the regression implementation closely. In the regression neural network, we have the covariates as inputs to the feedforward network. Here, we have the lagged values of the time series as input.

The use of artificial neural network applications in forecasting is surveyed in Hill et al. [88] and Zhang et al. [229]. Here we will describe some recent and non-standard applications.

One of the series analyzed by Raposo [173] was the famous airline passenger data of Box-Jenkins based on numbers of international passengers by month from 1949 to 1960. The model for this series becomes the standard benchmark model for seasonal data with trend: the multiplicative seasonal integrated moving average model also known as SARIMA $(0,1,1)_{12}(0,1,1)$.

Table 6.6 compares results for 1958 through 1960 in terms of mean average percentage error (MAPE) of the Box-Jenkins model and a neural network.

Table 6.6
Forecasting airline passenger loads.

Year	MAPE %			
	(12:2:1)	(13:2:1)	(14:2:1)	SARIMA
1958	7.63	6.44	5.00	7.75
1959	6.32	4.90	7.75	10.86
1960	7.39	7.93	11.91	15.55

(a:b:c) - number of neurons in: a - input, b - hidden, c - output, layer.

The airline data was also analyzed by Faraway and Chatfield [68] using a neural network.

Another classical time series examining yearly sunspot data was analyzed by Park et al. [159]. Their results are presented in Tables 6.7 and 6.8 for the training data (88 observations) and forecasting period (12 observations). Here they used the PCA method described in Section 2.8.

A multivariate time series forecasting using a neural network is presented in Chakraborthy et al. [34]. They used indices of monthly flour prices in Buffalo (x_t), Minneapolis (y_t) and Kansas City (z_t) over the period from August 1972 through November 1980, previously analyzed by Tiag and Tsay using an autoregressive integrated moving average (ARIMA) $(1,1)$ multivariate model. They used two classes

Table 6.7
Results from network structures and the AR(2) model.

Network Structure	Mean Square Error (MSE)	Mean Absolute Error[7] (MAE)
2:11:1	156.2792	9.8539
2:3:1	154.4879	9.4056
2:2:1*	151.9368	9.2865
2:1:1	230.2508	12.2454
2:0:1¨	233.8252	12.1660
AR(2)	222.9554	11.8726

* = Proposed optimal structure.
¨ = Network without hidden units.

Table 6.8
Summary of forecasting errors produced by neural network models and the AR(2) model based on untrained data.

Model	Mean Square Error (MSE)	Mean Absolute Error (MAE)
2:11:1	160.8409	9.2631
2:3:1	158.9064	9.1710
2:2:1*	157.7880	9.1522
2:1:1	180.7608	11.0378
2:0:1	180.4559	12.1861
AR(2)	177.2645	11.5624

* = Model selected by PCA method.

of feedforward networks: separate modeling of each series and a combined modeling as in Figures 6.8 and 6.9.

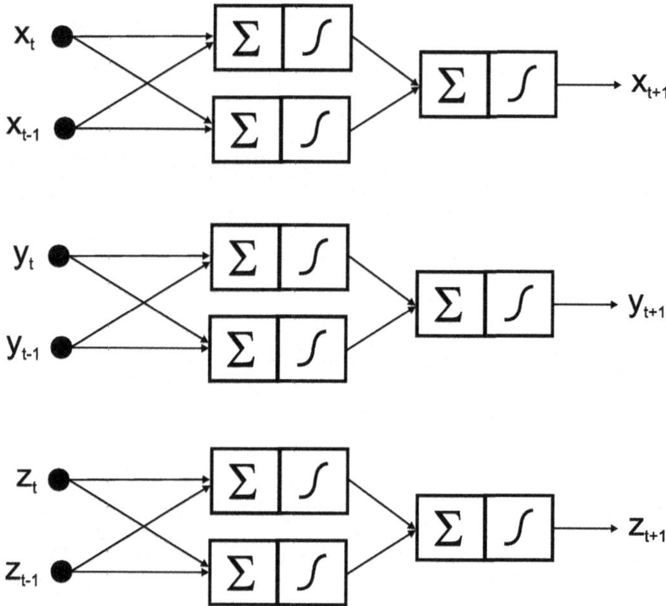

Figure 6.8
Separate architecture.

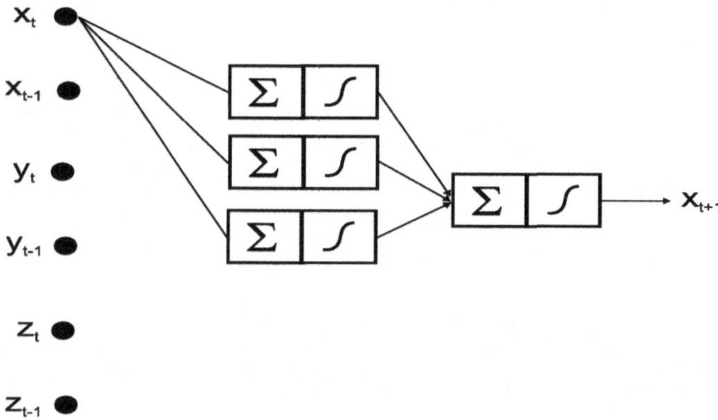

Figure 6.9
Combined architecture for forecasting x_{t+1} (and similar to y_{t+1} and z_{t+1}).

The results are shown in Tables 6.9 and 6.10.

In the previous example, initial analysis using true series methods have helped to design the architecture of the neural network. More explicit combinations of time series model and neural networks are: Tseng et al. [210] that improved the ARIMA

Table 6.9
Mean-squared errors (MSEs) for separate modeling and prediction $\times 10^3$.

Network		Buffalo	Minneapolis	Kansas City
		MSE	MSE	MSE
2-2-1	Training	3.398	3.174	3.526
	One-lag	4.441	4.169	4.318
	Multi-lag	4.483	5.003	5.909
4-4-1	Training	3.352	3.076	3.383
	One-lag	9.787	8.047	3.370
	Multi-lag	10.317	9.069	6.483
6-6-1	Training	2.774	2.835	1.633
	One-lag	12.102	9.655	8.872
	Multi-lag	17.646	14.909	13.776

Table 6.10
Mean-squared errors (MSEs) and coefficients of variation (CVs) for combined and Tiao and Tsay's modeling and prediction $\times 10^3$.

Model		Buffalo		Minneapolis		Kansas City	
		MSE	CV	MSE	CV	MSE	CV
6-6-1	Training	1.446	7.573	1.554	7.889	1.799	8.437
Network	One-lag	3.101	11.091	3.169	11.265	2.067	9.044
	Multi-lag	3.770	12.229	3.244	11.389	2.975	10.850
8-8-1	Training	0.103	2.021	0.090	1.898	0.383	3.892
Network	One-lag	0.087	1.857	0.072	1.697	1.353	7.316
	Multi-lag	0.107	2.059	0.070	1.674	1.521	7.757
Tiao/Tsay	Training	2.549	10.054	5.097	14.285	8.645	18.493
	One-lag	2.373	9.701	4.168	12.917	7.497	17.222
	Multi-lag	72.346	53.564	137.534	74.204	233.413	96.096

forecasting using a feedforward network with logged values of the series, its forecast and the residual obtained from the ARIMA fitting to the series; Calôba et al. [29] which used a neural network after decomposing the series from its low and high frequency signals; Cigizoglu [45] used generated data from a time series model to train a neural network. Periodicity terms and lagged values were used to forecast river flows in Turkey.

A comparison of structural time series model and neural networks is presented in Portugal [169]. Forecasting non-inversible time series using neural networks was attempted by Pino et al. [167] and combining forecasts by Donaldson and Kamstra [60]. Further comparisons with time series models and applications to classical time series data can be seen in Terasvirta et al. [205], Kajitani et al. [105], and Ghiassi et al. [76].

Modeling of nonlinear time series using a neural network and an extension called a stochastic neural network can be seen in Poli and Jones [168], Stern [199], Lai and Wong [114], and Shamseldin and O'Connor [191].

Forecasts using a combination of neural networks and fuzzy sets are presented by Machado et al. [135], Lourenço [130], and Li et al. [125].

Finally, an attempt to choose a classical forecast method using a neural network is shown in Venkatachalam and Sohl [214]. For automatic modeling with networks, see Balkin and Ord [14].

6.3 Control Chart Networks

Control charts are used to identify variations in production processes. There are usually two types of causes of variation: chance causes and assignable causes. Chance causes, or natural causes, are inherent in all processes. Assignable causes represent improperly adjustable machines, operation error, and defective raw materials and it is desirable to remove these causes from the process.

The process means (μ) and variance (σ^2) are estimated when observed data are collected in sample of size n and the means $\bar{X}_1, \bar{X}_2, \ldots$ are calculated. It is assumed that the sample means are mutually independent and that $\bar{X}_i$ has distribution $N(\mu_j, \frac{\sigma^2}{n})$, $i = 1, 2, \ldots$ and the condition $\mu = \mu_0$ is to be maintained. A shift in the process mean occurs when the distribution of the sample changes ($\mu \neq \mu_0$). The decision in the control chart is based on the sample mean with a signal of out-of-control being given at the first stage N such that

$$\delta_N = \frac{|\bar{X}_N - \mu_0|}{\sigma/\sqrt{n_0}} > c, \tag{6.19}$$

where usually $c = 3$ and the sample size is 4 or 5. This procedure is useful to detect shifts in the process mean. However, disturbances can also cause shifts or changes in the amount of process variability. A similar decision procedure can be used for the behavior of ranges of the samples from the process and will signal an out-of-control variability. Similarly chart for σ is usually based on

$$|\bar{R} - \sigma_R| > 3 \tag{6.20}$$

or

$$\bar{R} \pm 3A_m\bar{R}, \tag{6.21}$$

where $\bar{R}$ is average range of a number of sample ranges R and A_n is a constant which depends on n, relating σ to $E(R)$ and $V(R)$.

Several kinds of signals based in these statistics and charts have been proposed, for example: more than three consecutive values more than 2σ from the mean; eight consecutive values above the mean etc. This rules will detect trends, sinusoidal, and other patterns in the data.

These rules can also be implemented using neural networks, and we will now describe how some of these work. All of them use simulation results and make comparison with classical statistical charts in terms of percentage of wrong decisions and ARL (average run length) which is defined to be the expected number of samples that are observed from a process before out-of-control signal is signaled by the chart.

Prybutok et al. [170] used a feedforward network with 5 inputs (the sample size), a single hidden layer with 5 neurons and 1 output neuron. The activation functions were sigmoid; the neural network classifies the control chart as being either "in control" or "out-of-control" respectively according whether the output falls in the intervals $(0, 0.5)$ or $(0.5, 1)$.

Various rules were compared (see Prybutok et al. [170]) such as C_4=signal if eight of the last standardized sample means are between 0 and -3 or if the eight of the last eight are between 0 and 3. The rules were denoted by C_i and its combination by C_{ijkl}.

For eight shifts given by $|\mu - \mu_0|$ taking the values $(0, 0.2, 0.4, 0.6, 0.8, 1, 2, 3)$, the ratios of the ARL for the neural network to the standard Shewhart $\bar{X}$ control chart without supplemental runs rules are listed in Table 6.11. A further table with other rules appears in Prybutok et al. [170].

Smith [194] extends the previous work by considering a single model of a simultaneous X-bar and R chart, and investigating both location and variance shifts. She applied two groups of feedforward networks.

The first group utilized only one output, sigmoid activation function, two hidden layers, 3 or 10 neurons in each hidden layer, and interpretation as: 0 to 0.3 (mean shift), 0.3 to 0.7 (under control), 0.7 to 1 (variance shift).

The inputs were either in ten observations $(n = 10)$ and the calculated statistics (sample mean, range, standard deviation), a total of 13 inputs or on the calculated statistics only, i.e., 3 inputs. Table 6.12 shows the percentage of wrong decisions for the control chart and the neural networks.

The second group of networks differed from the first group by the input which now is the raw observations only $(n = 5, n = 10, n = 20)$. These neural networks were designed to recognize the following shapes: flat (in control), sine wave (cyclic), slope (trend of drift), and bimodal (up and down shifts). The networks here had two

Table 6.11
Ratios of average run lengths: neural network model Cn_{ij} to Shewhart control chart without runs rules (C_1).

Control	Shift (μ)							
Charts	0.0	0.2	0.4	0.6	0.8	1.0	2.0	3.0
ARL^*	370.40	308.43	200.08	119.67	71.55	43.89	6.30	2.00
C_1	1.00	1.00	1.00	1.00	1.00	1.00	1.00	1.00
Cn_{54}	0.13	0.14	0.16	0.18	0.21	0.24	0.27	0.19
Cn_{55}	0.25	0.27	0.30	0.33	0.38	0.38	0.46	0.32
Cn_{56}	0.47	0.55	0.60	0.67	0.68	0.72	0.71	0.44
Cn_{57}	0.95	0.99	1.31	1.41	1.48	1.44	1.19	0.68
Cn_{62}	0.18	0.19	0.22	0.25	0.30	0.32	0.43	0.34
Cn_{63}	0.24	0.26	0.29	0.32	0.37	0.43	0.50	0.36
Cn_{64}	0.33	0.34	0.36	0.39	0.44	0.48	0.55	0.43
Cn_{65}	0.42	0.40	0.44	0.48	0.53	0.56	0.67	0.53
Cn_{66}	0.56	0.53	0.57	0.60	0.63	0.69	0.76	0.56
Cn_{67}	0.78	0.71	0.72	0.78	0.82	0.79	0.90	0.63
Cn_{68}	1.15	1.02	0.94	1.01	1.02	1.08	1.04	0.75
Cn_{71}	0.25	0.26	0.27	0.33	0.33	0.38	0.44	0.07
Cn_{72}	0.49	0.52	0.52	0.60	0.61	0.63	0.70	0.42
Cn_{73}	0.88	0.92	0.96	1.01	1.02	1.09	0.98	0.63
Cn_{74}	1.56	1.61	1.71	1.72	1.69	1.56	1.47	0.90

$ARL = ARL_{c_1}(\mu)$ for Shewhart Control Chart.

Table 6.12
Wrong decisions of control charts and neural networks.

Training/ Test Set	Inputs	Hidden Layer Size	Neural Net Percent Wrong	Shewhart Control Percent Wrong
A	All	3	16.8	27.5
A	All	10	11.7	27.5
A	Statistics	3	28.5	27.5
B	All	3	0.8	0.5
B	All	10	0.5	0.5
B	Statistics	3	0.0(None)	0.5

outputs used to classify patterns with the codes: flat $(0,0)$, sine wave $(1,1)$, trend $(0,1)$, and bimodal $(1,0)$. A output in $(0,0.4]$ was considered 0 and in $(0.6,1)$ was considered 1. Table 6.13 gives some of the results Smith [194] obtained.

Hwarng [93] presents a neural network methodology for monitoring process shift in the presence of autocorrelation. The study of AR (1) (autoregressive of order one) processes shows that the performance of the neural network based monitoring

Table 6.13
Desired output vectors.

Pattern	Desired output vector	Note
Upward trend	[1,0,0,0,0]	–
Downward trend	[-1,0,0,0,0]	–
Systematic variation (I)	[0,1,0,0,0]	First observation is above the in-control mean
Systematic variation (II)	[0,-1,0,0,0]	First observation is below the in-control mean
Cycle (I)	[0,0,1,0,0]	Sine wave
Cycle (II)	[0,0,-1,0,0]	Cosine wave
Mixture (I)	[0,0,0,1,0]	Mean of first distribution is greater than in-control mean
Mixture (II)	[0,0,0,-1,0]	Mean of first distribution is less than in-control mean
Upward shift	[0,0,0,0,1]	–
Downward shift	[0,0,0,0,-1]	–

scheme is superior to the other five control charts in the cases investigated. He used a feedforward network with $50 \cdot 30 \cdot 1$ (input·hidden·output) structure of neurons.

A more sophisticated pattern recognition process control chart using neural networks is given in Cheng [37]. A comparison between the performance of a feedforward network and a mixture of expert was made using simulations. The multilayer network had 16 inputs (raw observations), 12 neurons in the hidden layers and 5 outputs identifying the patterns as given in Table 6.14. Hyperbolic activation function was used.

A mixture of three experts (choice based on experimentation) was used with the same input as in the feedforward network. Each expert had the same structure as the feedforward network. The gating network has 3 neurons elements in the output layer and 4 neurons in a hidden layer.

The results seem to indicate that the mixture of experts performs better in terms of correctly identified patterns.

6.4 Some Statistical Inference Results

In this section, we describe some inferential procedures that have been developed for using neural networks. In particular we outline some results on point estimation

Table 6.14
Pattern recognition networks.

# Input Points	σ of Noise	# Correct of 300	# Missed of 300			
			Flat	Slope	Sine wave	Bimodal
5	0.1	288	0	0	10	2
5	0.2	226	39	15	10	10
5	0.3	208	32	39	18	3
10	0.1	292	0	0	5	3
10	0.2	268	3	7	16	6
10	0.3	246	9	17	23	5
20	0.1	297	0	0	0	3
20	0.2	293	0	0	0	7
20	0.3	280	6	1	1	12
Over All Networks		88.8%	3.3%	2.9%	3.1%	1.9%

(Bayes, likelihood, robust), interval estimation (parametric, bootstrap), and significance tests (F-test and nonlinearity test).

For a more detailed account of the prediction interval outlined here see also Hwang and Ding [92].

6.4.1 Estimation Methods

The standard method of fitting a neural network is *backpropagation*, which is a gradient descent algorithm to minimize the sum of squares. Here we outline some details of other alternative methods of estimation.

Since the most used activation function is the sigmoid we present some further results. First consider, maximum likelihood estimation; we follow Schumacher et al. [189] and Faraggi and Simon [65].

The logistic (sigmoid) activation relates output y to input x assuming

$$P_{\boldsymbol{x}}(Y = 1/\boldsymbol{x}) = \wedge\left(\omega_0 + \sum_{i=1}^{I} \omega_i x_{ji}\right) = p(\boldsymbol{x}, \omega), \tag{6.22}$$

where $\wedge(u) = 1/(1 + e^u)$ is the sigmoid or logistic function.

As seen in Section 5.1, this activation function as a nonlinear regression model is a special case of the generalized linear model, i.e.,

$$E(Y/x) = p(x, \omega), \;\; V(Y/\boldsymbol{x}) = p(x, \omega)(1 - p(x, \omega)). \tag{6.23}$$

The maximum likelihood estimator of the weights ω's is obtained from the log likelihood function

$$L(\omega) = \sum_{j=1}^{n} \left[y_j \log p(x_i, \underset{\sim}{\omega}) + (1 - y_j) \log(1 - p(x_j \omega) \right], \tag{6.24}$$

where (x_j, y_j) is the observation for pattern j.

To maximize $L(\omega)$ is equivalent to minimize the Kulback-Leiber divergence

$$\sum_{j=1}^{n} \left[y_j \log \frac{y_j}{p(x_j \omega)} + (1 - y_j) \log \frac{1 - y_j}{1 - p(x_j, \omega)} \right]. \tag{6.25}$$

It can be written as

$$\sum_{j=1}^{n} - \log \left(1 - |y_j - p(x, \omega)|\right), \tag{6.26}$$

which makes the interpretation of distance between y and $p(y, \omega)$ more clear. At this point, for comparison, we recall the criteria function used in the backpropagation least square estimation method (or learning rule)

$$\sum_{j=1}^{n} \left(y_j - p(x_j, \omega)\right)^2. \tag{6.27}$$

Due to the relation to of maximum likelihood criteria and the Kulback-Leiber, the estimation based on Equation (6.25) is called backpropagation of maximum likelihood. In the neural network, jargon it is also called relative entropy or cross entropy method.

6.4.2 Bayesian Methods

Bayesian methods for neural networks rely mostly in MCMC (Markov Chain Monte Carlo). Here, we follow Lee [122] and review the priors used for neural networks.

All approaches start with the same basic model for the output y

$$y_i = \beta_0 + \sum_{j=1}^{k} \beta_j \frac{1}{1 + \exp(-\omega_{j0} - \sum_{h=1}^{p} \omega_{jh} x_{ih})} + \epsilon_i, \tag{6.28}$$

$$\epsilon_i \sim N(0, \sigma^2). \tag{6.29}$$

The Bayesian models are:

- **Insua and Müller** (1998)
  A direct acyclic graph (DAG) diagram of the model is in Figure 6.10.

  The distributions for parameters and hyperparameters are:

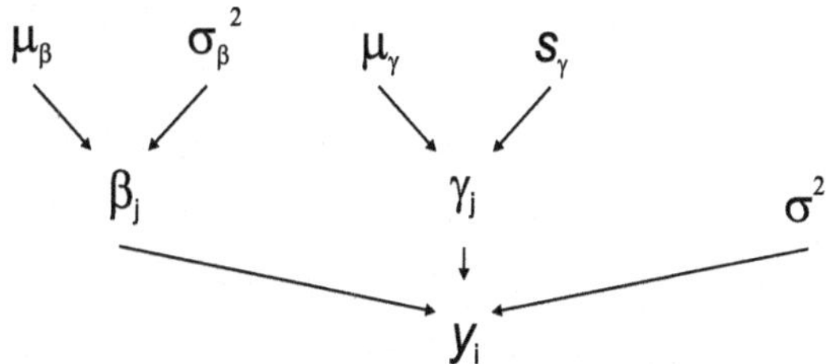

Figure 6.10
Graphical model of Muller and Rios Issua.

$$y_i \sim N\left(\sum_{j=0}^{m} \beta_j \wedge (\omega_j' x_i), \sigma^2\right), \; i = 1, \ldots, N \tag{6.30}$$

$$\beta_j \sim N(\mu_\beta, \sigma_\beta^2), \; j = 0, \ldots, m \tag{6.31}$$

$$\omega_j - \omega\gamma_j \sim N_p(\mu_\gamma, \mathbf{S}_\gamma), \; j = 1, \ldots, m \tag{6.32}$$

$$\mu_\beta \sim N(a_\beta, A_\beta) \tag{6.33}$$

$$\mu_\gamma \sim N_p(a_\gamma, \mathbf{A}_\gamma) \tag{6.34}$$

$$\sigma_\beta^2 \sim \Gamma^{-1}\left(\frac{c_\beta}{2}, \frac{C_\beta}{2}\right) \tag{6.35}$$

$$\mathbf{S}_\gamma \sim Wish^{-1}\left(c_\gamma, (c_\gamma \mathbf{C}_\gamma)^{-1}\right) \tag{6.36}$$

$$\sigma^2 \sim \Gamma^{-1}\left(\frac{s}{2}, \frac{S}{2}\right). \tag{6.37}$$

There are fixed hyperparameters that need to be specified. Insua and Müller [97] specify many of them to be of the same scale as the data. As some fitting algorithms work better when (or sometimes only when) the data have been re-scaled so that $|x_{ih}|, |y_i| \leq 1$, one choice of starting hyperparameters is: $a_\beta = 0$, $A_\beta = 1$, $a_\gamma = \mathbf{0}$, $\mathbf{A}_\gamma = \mathbf{I}_p$, $c_\beta = 1$, $C_\beta = 1$, $c_\gamma = p + 1$, $\mathbf{C}_\gamma = \frac{1}{p+1}\mathbf{I}_p$, $s = 1$, $S = \frac{1}{10}$.

- **Neal Model** (1996)
  Neal's four-stage model contains more parameters than the previous model. A DAG diagram of the model is in Figure 6.11.

  Each of the original parameters (β and γ) is treated as a univariate normal with mean zero and its own standard deviation. These standard deviations are the product of two hyperparameters, one for the originating node of the link in the graph, and one for the destination node.

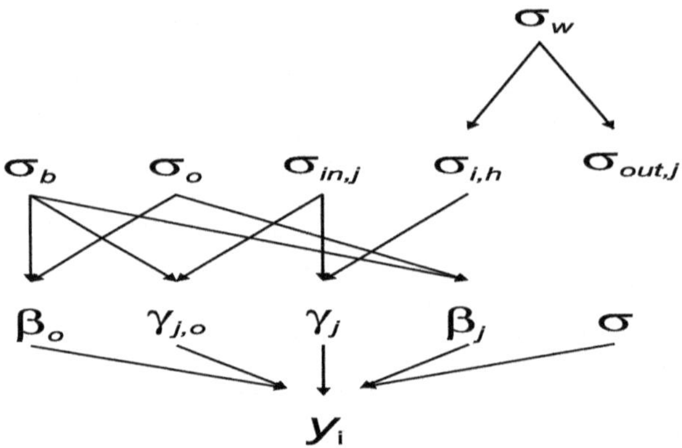

Figure 6.11
Graphical model of Neal.

There are two notes on this model that should be mentioned. First, Neal [149] uses hyperbolic tangent activation functions rather than logistic functions. These are essentially equivalent in terms of the neural network, the main difference being that their range is from -1 to 1 rather than 0 to 1. The second note is that Neal also discusses using t distributions instead of normal distributions for the parameters.

- **MacKay Model** (1992)
MacKay starts with the idea of using an improper prior, but knowing that the posterior would be improper, he uses the data to fix the hyperparameters at a single value. Under his model, the distributions for parameters and hyperparameters are:

$$y_i \sim N\left(\sum_{j=0}^{m} \beta_j \wedge (\omega_j' x_i)(\gamma_j' x_i), \frac{1}{\nu}\right), \ i = 1, \ldots, N \tag{6.38}$$

$$\beta_j \sim N\left(0, \frac{1}{\alpha_1}\right), \ j = 0, \ldots, m \tag{6.39}$$

$$\omega_j h \sim N_p\left(0, \frac{1}{\alpha_2}\right), \ j = 1, \ldots, m \ h = 1, \ldots, p \tag{6.40}$$

$$\omega_j 0 \sim N_p\left(0, \frac{1}{\alpha_3}\right), \ j = 1, \ldots, m \tag{6.41}$$

$$\alpha_k \propto 1, \ k = 1, 2, 3 \tag{6.42}$$

$$\upsilon \propto 1. \tag{6.43}$$

Mackay [136] uses the data to find the posterior mode for the hyperparametrics α and ν and then fixes them at their modes.

Each of the three models reviewed is fairly complicated and was fit using MCMC.

- **Lee Model** (1998)
The model for the response y is

$$y_i = \beta_0 + \sum_{j=1}^{k} \beta_j \frac{1}{1 + \exp(-\gamma_{j0} - \sum_{h=1}^{p} \omega_{jh} x_{ih})} + \epsilon_i, \tag{6.44}$$

$$\epsilon_i \overset{iid}{\sim} N(0, \sigma^2).$$

Instead of having to specify a proper prior, Lee [118] uses the noninformative improper prior

$$\pi(\beta, \omega, \sigma^2) = (2\pi)^{-d/2} \frac{1}{\sigma^2}, \tag{6.45}$$

where d is the dimension of the model (number of parameters). This prior is proportional to the standard noninformative prior for linear regression. This model is also fitted using MCMC.

This subsection ends with some references on estimation problems of interest. Belue and Bauer [16] give some experimental design procedures to achieve higher accuracy on the estimation of the neural network parameters. Copobianco [49] outlines robust estimation framework for neural networks. Aitkin and Foxall [6] reformulate the feedforward neural network as a latent variable model and construct the likelihood, which is maximized using finite mixture and the EM (expectation-maximization) algorithm.

6.4.3 Interval Estimation

Interval estimation for the feedforward neural network essentially uses the standard results of least squares for nonlinear regression models and ridge regression (regularization). Here we summarize the results of Chryssolouriz et al. [40] and De Veaux et al. [55].

Essentially, they used the standard results for the non-linear regression least square method

$$S(\omega) = \sum (y_i - f(X, \omega))^2 \tag{6.46}$$

or regularization, when weight decay is used

$$S(\omega) = \sum (y_i - f(X, \omega))^2 + k\alpha \sum \omega_i^2. \tag{6.47}$$

The corresponding prediction intervals are obtained from the t distribution and

$$t_{n-p} s \sqrt{1 + g_0'(JJ)^{-1} g_0} \tag{6.48}$$

$$t_{n-p^*} s^* \sqrt{1 + g_0'(J'J + kI)^{-1}(J'J + kI)^{-1} g_0} \tag{6.49}$$

where $s =$ (residual sum of squares)$/(n-p) = RSS/(n-p)$, p is the number of parameters, J is a matrix with entries $\partial f(\omega, x_i)/\partial \omega_j$ g_0 is a vector with entries $\partial f(x_0 \omega)/\partial \omega_j$ and

$$k = y - f(X\omega^*) + J\omega^*, \tag{6.50}$$

$$s^* = RSS/(n - tr(2H - H^2)) \tag{6.51}$$

$$H = J(J'J + \alpha I)^{-1} J \tag{6.52}$$

$$p^* = tr(2H - H^2), \tag{6.53}$$

where $H = J(J'J + \alpha I)^{-1} J'$, ω^* is the true parameter value.

The author gives an application and De Veaux et al. [55] also show some simulated results.

Tibshirani [207] compared seven methods o estimate prediction intervals for neural networks: (1) delta method; (2) approximate delta using approximation that ignores second derivatives for information matrix; (3) regularization estimator; (4) and (5) approximate sandwich estimators; (6) and (7) bootstrap estimators.

In these examples and simulations, Tibshirani found that the bootstrap methods provided the most accurate estimates of the standard errors of the predicted values. The non-simulation methods (delta, regularization, sandwich estimators) missed the variability due to random choice of starting values.

6.4.4 Statistical Tests

Two statistical tests using neural networks are presented here.

Adams [3] used the following procedure to test a hypothesis about the parameters (weights) of a neural network. The test can be used to test nonlinear relations between the inputs and outputs variables. To implement the test, it is first necessary to calculate the degree of freedom for the network.

If n is the number of observations and p the number of estimated parameters, then the degree of freedom (df) is calculated as

$$df = n - p. \tag{6.54}$$

With a multilayer feedforward network with m hidden layers, the input layer designated by 0 and the out designated by $m+1$. For L_{m+1} outputs and L_m neurons in the hidden layer we have

$$df = n - p + (L_{m+1} - 1)(L_n + 1). \tag{6.55}$$

The number of parameters is

$$P = \sum_{i=1}^{m+1} \left(L_{i-1} L_1 + L_i \right). \tag{6.56}$$

For a network with 4:6:2, i.e. 4 inputs, 6 hidden neurons and 2 outputs, the degree of freedom for each output is

$$df = n - 44 + (2 - 1)(6 + 1) = n - 37. \tag{6.57}$$

For a feedforward network trained by least square we can form the model sum of squares. Therefore, any submodel can be tested using the F test

$$F = \frac{(SSE_0 - SSE)/(df_0 - df)}{SSE/df}, \tag{6.58}$$

where SSE_0 is the sum of squares of submodel and SSE is the sum of squares of full model.

The above test is based on the likelihood ratio test. An alternative proposed by Lee et al. [121] uses the Rao score test or Lagrange multiplier test. Blake and Kapetanios [21] also suggested a Wald equivalent test. See also Lee [120].

The implementation of the Rao test (see Kiani [110]) to test nonlinearity of a time series using a one-hidden layer neural network is

(i) Regress y_t on intercept and $y_{t-1}, \ldots, y_{t-k}$, save residuals $(\hat{u}_t)$, residual sum of squares SSE_1 and predictions $\hat{y}_t$,

(ii) Regress $\hat{u}_t$ on $y_{t-1}, \ldots, y_{t-k}$ using neural network model that nests a linear regression $y_t = \pi \underset{\sim t}{y} + u_t(\underset{\sim t}{y} = (y_{t-1}, \ldots, y_{t-k})$, save matrices Ψ of outputs of the neurons for principal components analysis.

(iii) Regress $\hat{e}_t$ (residuals of the linear regression in (ii)) in Ψ^* (principal components of Ψ) and X matrices, and obtain R^2, thus

$$TS = nR^2 \sim \chi^2(g) \tag{6.59}$$

where g is the number of principal components used un step (iii).

The neural network in Figure 6.12 clarifies the procedure.

Simulation results on the behaviors of this test can be seen in the bibliography. An application is described in Kiani [110] and power properties are discussed in Terasvirta et al. [206].

6.5 Forecasting with Python

In this section, we use a dataset containing the pandas library of Python to predict stock market fluctuations. Similar models can be used to predict house prices and

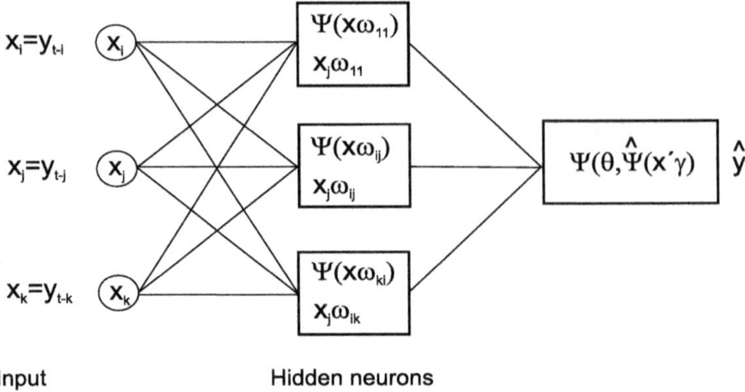

Figure 6.12
One-hidden layer feedforward neural network.

any other time series whose previous values are related to further values. You can use the scikit-survival library of Python for the Cox proportional model. For the control chart, you can use radial basis function networks, see Addeh et al. [4].

The example given needs the dataset `stocks.csv` that is available at the address `http://www.lncc.br/~borges/Statistical-Learning/` with supplemental material for this book. Figure 6.13 depicts the data. The time is on the x axis, and the stock market closing prices is on the y axis. The dataset has one column with dates and another with values as follows.

```
        Date   Close
0  4/25/2011  15.77
1  4/26/2011  15.62
2  4/27/2011  15.35
3  4/28/2011  14.62
4  4/29/2011  14.75
```

You can use any dataset with dates and values. In addition, you can create new datasets using a spreadsheet for saving the data as a comma-separated-values (CSV) file.

Considering that we want to test the prediction model with the existent data, we need to split the dataset into data for training the neural network and into data for testing the prediction. Figure 6.14 depicts the data used for training on the left hand and the data used for testing on the right hand.

Note that a specific label for a date is not related with a share price. The way

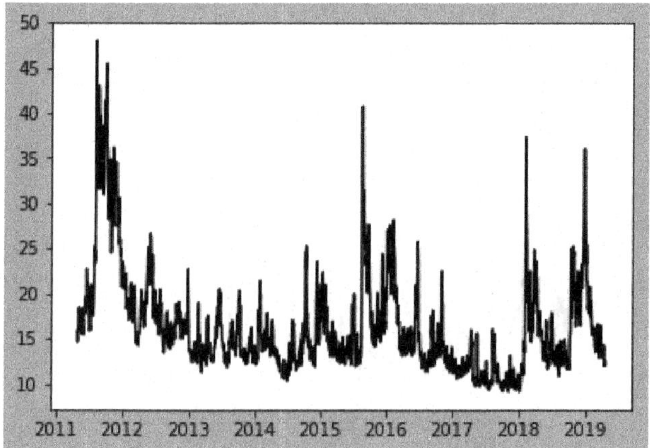

Figure 6.13
Stock market.

that we label the time is not relevant. Thus, whether the calendar is lunar, solar, or lunisolar is not relevant. Hence, we cannot use the date as an input for the network. However, the earlier data collected is relevant, and we need both axes to find a plane curve, i.e., we need x and y. In fact, time series is a sequence of data points collected over time. The big insight is how to set the relation between x and y. If s is the sequence, we need to set

$$y_i \leftarrow s_i \qquad \text{or} \qquad y_i = s_i,$$

and then

$$x_i \leftarrow s_{i-1} \qquad \text{or} \qquad x_i = s_{i-1},$$

for every i in the sequence s. Considering the ordered pair (x_i, y_i), the sequences x and y should have the same number of terms. Consequently, we lose the first term of the sequence in y and the last term of the sequence in x, and such sequences contain one term less than sequence s. Program 6.1 sets sequences in lines 26 to 29. Line 6 reads the dataset, and line 8 prints its beginning. Lines 10 and 11 index the dataset in time, and line 12 shows the differences. Lines 14 and 15 generate the Figure 6.13. Lines from 17 to 20 split the data into train and test. Lines from 22 to 25 generates the Figure 6.14. Lines from 36 to 38 only set the dates for the Figure 6.15 generated in lines from 40 to 45.

Program 6.1
Stock market forecast.

```
1 import pandas as pd
2 import matplotlib.pyplot as plt
3 from sklearn.neural_network import MLPRegressor
4 from datetime import timedelta
```

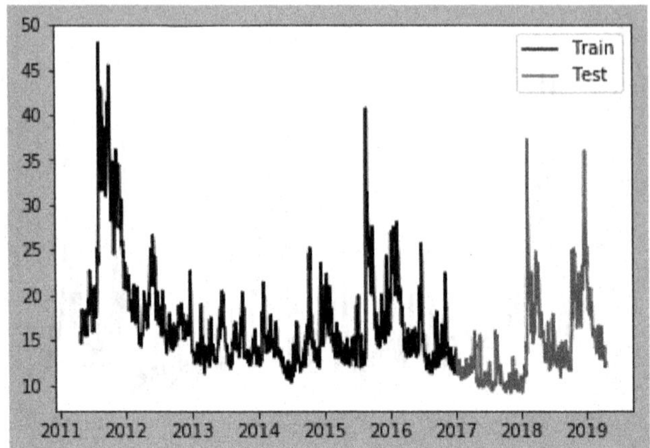

Figure 6.14
Stock market data split into training and test data.

```
plt.style.use('grayscale')

df = pd.read_csv("stocks.csv")
print(df.head())

df['Date'] = pd.to_datetime(df['Date'])
df = df.set_index(['Date'])
print(df.head())

plt.plot(df['Close']);
plt.show()

split_date = pd.Timestamp('2017-01-01')
df = df['Close']
train = df.loc[:split_date]
test = df.loc[split_date:]

plt.plot(train)
plt.plot(test)
plt.legend(['Train', 'Test']);
plt.show()

X_train = train.values.reshape(-1, 1)[:-1]
y_train = train.values[1:]
X_test = test.values.reshape(-1, 1)[:-1]
y_test = test.values[1:]

```

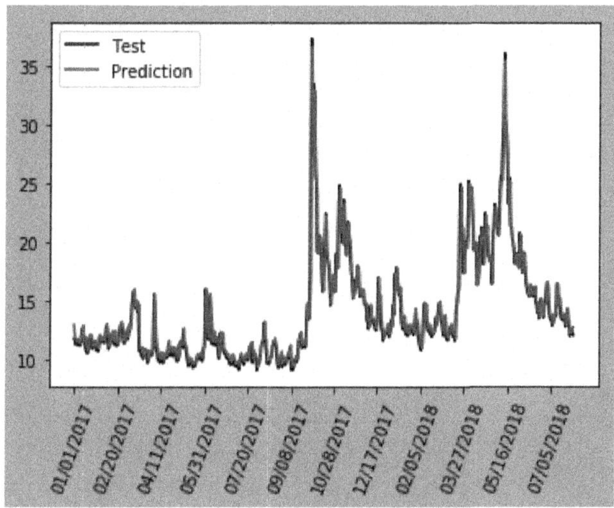

Figure 6.15
Stock market forecast.

```
NN = MLPRegressor()
n = NN.fit(X_train, y_train)
y_predicted = NN.predict(X_test)

dates = []
for i in range(0,len(y_test),50):
    dates.append((split_date + timedelta(days=i)).strftime('%m/%d/%Y'))

plt.plot(y_test, label='Test')
plt.plot(y_predicted, label='Prediction',)
plt.xticks(range(0,len(y_test),50),dates)
plt.xticks(rotation=70)
plt.legend()
plt.show();
```

Usually, the neural network models scale their data between −1 and 1 to obtain better results in the prediction. It can be done using the command `MinMaxScaler` from `sklearn.preprocessin`. A better prediction model can be written with a recurrent neural network known as LSTM (long short-term memory; see Hochreiter and Schmidhuber [89]).

Now, you have the foundations to understand neural network models and their relations with statistics. Moreover, you have the basis to teach yourself and search new commands and libraries to implement your own models in Python. We wish you success in your efforts.

A

Command Reference

This appendix presents a reference for internal and external Python commands in alphabetic order with several examples for the command-line interface.

Internal Commands

We present some commands in the Python Standard Library, i.e., the built-in commands.

def *function*(*parameters*)

defines a new *function* with its *parameters*.

```
>>> def prime(n):
...     if n in [2,3,5] or 2**n%n==2 and 3**n%n==3 and 5**n%n==5:
...             print("I guess %d is a prime number." %(n))
...     else:
...             print("I guess %d is not a prime number." %(n))
...
>>> prime(7)
I guess 7 is a prime number.
>>> prime(2)
I guess 2 is a prime number.
>>> prime(101)
I guess 101 is a prime number.
>>> prime(102)
I guess 102 is not a prime number.
>>>
```

for *i* in *object*:

iterates *i* for each *iterable* object. This command requires indentation for the block of commands to be iterated.

```
>>> t=""
>>> s="Statistics"
```

```
>>> for i in s:
...     t=t+i
...     print(t)
...
S
St
Sta
Stat
Stati
Statis
Statist
Statisti
Statistic
Statistics
>>>
>>> t=""
>>> for i in range(5):
...     for j in range(5):
...             t=t+str(i)+str(j)
...     print(t)
...     t=""
...
0001020304
1011121314
2021222324
3031323334
4041424344
>>>
```

if *c*:

runs a block of commands if the logical condition *c* is satisfied. The block must be indented to mark the block.

```
>>> a=1
>>> b=2
>>>
>>> if a<b:
...     if a!=b:
...             if a==a or b==b:
...                     print("Conditions fulfilled!")
...
...
Conditions fulfilled!
```

```
>>>
>>> if (a>b)^(a<b): print("Either a>b or a<b")
...
Either a>b or a<b
>>> if a==b:
...       print("Equal")
... else:
...       print("Different")
...
Different
>>>
```

import *module*

loads modules from Python code, like a library in other languages. When Python runs an import, the programmer can use the commands in the *module*. You can import a specific *resource* using the syntax `from module import resource`.

```
>>> numpy.sqrt(2)
Traceback (most recent call last):
File "<stdin>", line 1, in <module>
NameError: name 'numpy' is not defined
>>> import numpy
>>> numpy.sqrt(2)
1.4142135623730951
>>> sqrt(2)
Traceback (most recent call last):
File "<stdin>", line 1, in <module>
NameError: name 'sqrt' is not defined
>>> from numpy import sqrt
>>> sqrt(2)
1.4142135623730951
>>>
```

int(*string*, *base*)

truncates a number and converts a *string* written in a *base* into an integer in decimal, for instance, the three commands `int("9")`, `int(9.9)`, and `int("1001", 2)` return the same value, namely, 9.

```
>>> int("9")
9
```

```
>>> int(9.9)
9
>>> int("1001",2)
9
>>>
```

len(*O*)
returns the number of elements in an object *O*.

```
>>> a={42,51,13,11}
>>> len(a)
4
>>> b='Hello World!'
>>> len(b)
12
>>> c="Hello World!"
>>> len(c)
12
>>> a
set([11, 42, 51, 13])
>>> b
'Hello World!'
>>> c
'Hello World!'
>>>
```

max()
either returns the largest parameter or returns the largest element.

```
>>> max(42,51,13,11)
51
>>> a={42,51,13,11}
>>> max(a)
51
>>>
```

min()
either returns the smallest parameter or returns the smallest element.

```
>>> min(42,51,13,11)
11
>>> a={42,51,13,11}
>>> min(a)
11
>>>
```

print(*messages*)

prints *messages* onto the screen, for instance, `print( ("apple", 45, "banana") )` will print `("apple", 45, "banana")`.

```
>>> print( ("apple", 45, "banana") )
('apple', 45, 'banana')
>>> print("Hello World!")
Hello World!
>>>
```

quit()

ceases to interpret the Python commands.

```
MacBook:~ user$ python
>>> quit()
MacBook:~ user$ exit
logout
Saving session...
...copying shared history...
...saving history...truncating history files...
...completed.
Deleting expired sessions...6 completed.

[Process completed]
```

range(*start, stop, step*)

returns an iterable object with $\frac{stop-start}{step}$ elements from *start* to *stop* and interval *step*. If we use only the parameter *stop*, the sequence goes from zero to *stop - 1*.

```
>>> range(5)
[0, 1, 2, 3, 4]
```

```
>>> range(1,6,2)
[1, 3, 5]
>>>
```

str(*n*)

converts the number *n* into a string.

```
>>> str(9)
'9'
>>> str(9.9)
'9.9'
>>>
```

External Commands

We present some commands from other libraries, not built-in commands. To run them, we might need to install the library, and we need to run the internal command `import`.

datetime.timedelta(days=*i*)

returns the date after *i* days.

```
>>> import datetime
>>> a=datetime.date.today()
>>> b=datetime.timedelta(days=5)
>>> a
datetime.date(2019, 11, 11)
>>> b
datetime.timedelta(days=5)
>>> a+b
datetime.date(2019, 11, 16)
>>>
```

matplotlib.gridspec.GridSpec(*r*, *c*)

returns a grid with *r* rows and *c* columns to positioning figures.

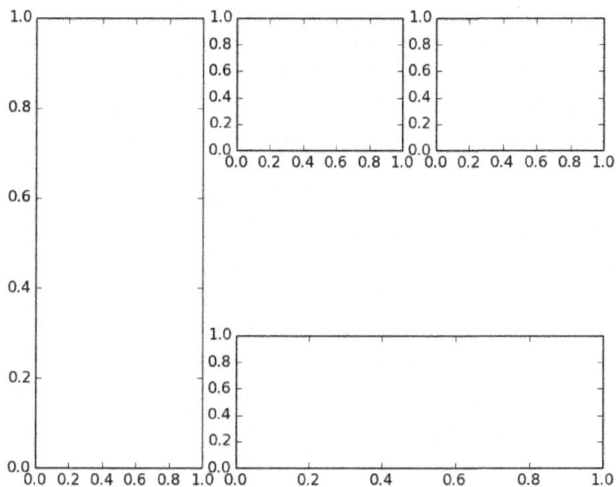

Figure A.1
matplotlib.gridspec.GridSpec

```
>>> import matplotlib.gridspec
>>> import matplotlib.pyplot
>>> G=matplotlib.gridspec.GridSpec(3,3)
>>> matplotlib.pyplot.subplot(G[0:,0])
<matplotlib.axes.AxesSubplot object at 0x113a93d50>
>>> matplotlib.pyplot.subplot(G[0,1])
<matplotlib.axes.AxesSubplot object at 0x11514f190>
>>> matplotlib.pyplot.subplot(G[0,2])
<matplotlib.axes.AxesSubplot object at 0x113ee2310>
>>> matplotlib.pyplot.subplot(G[2,1:])
<matplotlib.axes.AxesSubplot object at 0x113b88e10>
>>> matplotlib.pyplot.show()
>>>
```

The example generates the Figure A.1.

matplotlib.pyplot.axis('off')
turns off axes in a plot.

```
>>> import matplotlib.pyplot
>>> matplotlib.pyplot.axis('off')
```

```
(0.0, 1.0, 0.0, 1.0)
>>> matplotlib.pyplot.show()
>>> matplotlib.pyplot.axis('on')
(0.0, 1.0, 0.0, 1.0)
>>> matplotlib.pyplot.show()
>>>
```

matplotlib.pyplot.contourf(*X, Y, Z, levels=[* $l_1, \ldots$ *]*)
prepares a contour plot of the arrays X, Y, and Z that determine the coordinates, where *level* determines the number of the contour lines and their positions.

```
>>> import matplotlib.pyplot
>>> import numpy
>>> x=numpy.linspace(0,3,10)
>>> y=numpy.linspace(0,3,10)
>>> X,Y=numpy.meshgrid(x,y)
>>> Z=X*Y
>>> matplotlib.pyplot.contourf(X,Y,Z,cmap='gray_r')
<matplotlib.contour.QuadContourSet instance at 0x10b181a28>
>>> matplotlib.pyplot.show()
>>>
```

The example generates the Figure A.2.

matplotlib.pyplot.figure(figsize = (*w, h*))
creates a new figure with width *w* and height *h* in inches.

```
>>> import matplotlib.pyplot
>>> matplotlib.pyplot.figure(figsize=(2,5))
<matplotlib.figure.Figure object at 0x106355e90>
>>> matplotlib.pyplot.show()
>>>
```

matplotlib.pyplot.hist(*array, density=True, alpha=f*)
prepares a histogram of the *array*, where *density=True* requires that its area will sum to 1, and *alpha* is a float *f* from 0 (transparent image) to 1 (opaque image). Figure A.3 shows the histogram.

```
>>> import matplotlib.pyplot
>>> matplotlib.pyplot.hist([1,3,1,2,1,2],color='gray')
(array([3., 0., 0., 0., 0., 2., 0., 0., 0., 1.]),
```

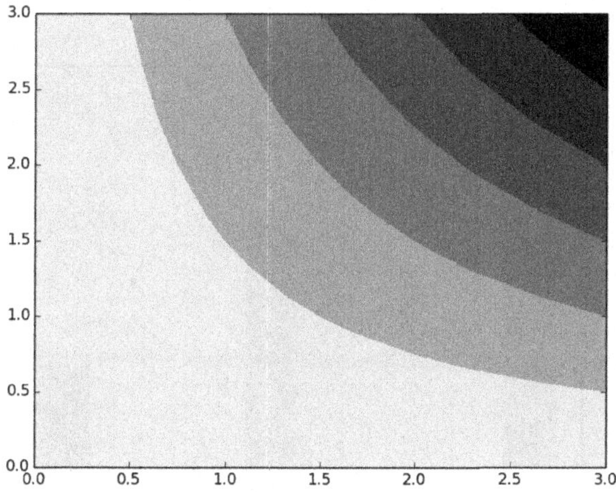

Figure A.2
matplotlib.pyplot.contourf

```
array([1., 1.2, 1.4, 1.6, 1.8,2., 2.2, 2.4, 2.6, 2.8, 3.]),
<a list of 10 Patch objects>)
>>> matplotlib.pyplot.show()
>>>
```

matplotlib.pyplot.imshow(*G*)
prepares a figure with a raster graphic *G*.

```
>>> import matplotlib.pyplot
>>> import numpy
>>> a=numpy.array([[1,2,3],[4,5,6],[7,8,9]])
>>> matplotlib.pyplot.imshow(a,cmap='gray')
<matplotlib.image.AxesImage object at 0x11122c990>
>>> matplotlib.pyplot.show()
>>>
```

The example generates the Figure A.4.

matplotlib.pyplot.legend()
prepares a legend for a plot.

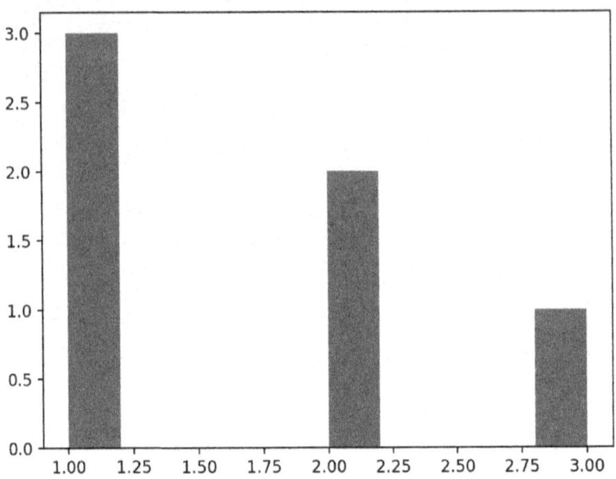

Figure A.3
matplotlib.pyplot.hist

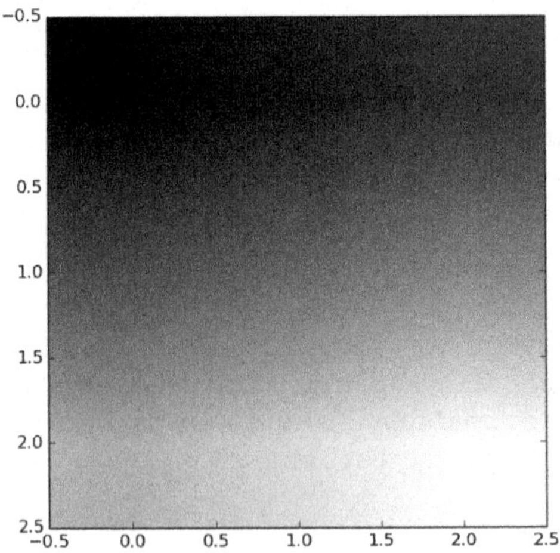

Figure A.4
matplotlib.pyplot.imshow

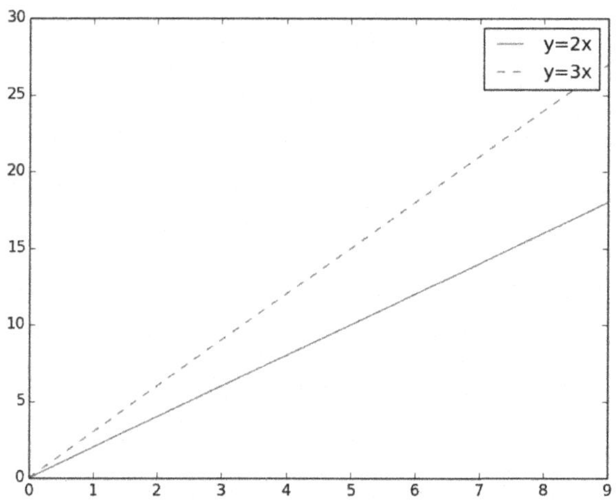

Figure A.5
matplotlib.pyplot.legend

```
>>> import matplotlib.pyplot
>>> import numpy
>>> x=numpy.arange(10)
>>> y1=2*x
>>> y2=3*x
>>> matplotlib.pyplot.plot(x,y1,'-',label='y=2x',c='gray')
[<matplotlib.lines.Line2D object at 0x1128196d0>]
>>> matplotlib.pyplot.plot(x,y2,'--',label='y=3x',c='gray')
[<matplotlib.lines.Line2D object at 0x112819190>]
>>> matplotlib.pyplot.legend()
<matplotlib.legend.Legend object at 0x113065ad0>
>>> matplotlib.pyplot.show()
>>>
```

The example generates the Figure A.5.

matplotlib.pyplot.plot(X,Y)
prepares the plot with the coordinates (x_i, y_i) such that $x_i \in X$ and $y_i \in Y$.

```
>>> import matplotlib.pyplot
>>> matplotlib.pyplot.plot([1,3,2],[5,4,6],color='grey')
[<matplotlib.lines.Line2D object at 0x121f80438>]
>>> matplotlib.pyplot.show()
```

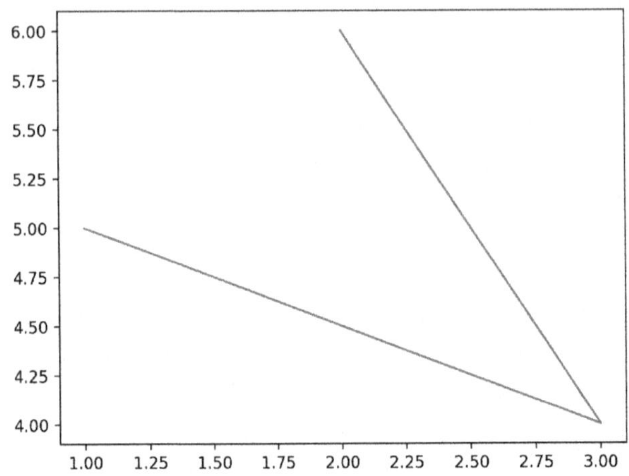

Figure A.6
matplotlib.pyplot.plot

It generates the Figure A.6.

matplotlib.pyplot.scatter(X, Y, marker='D')
prepares the scatter plot with the coordinates (x_i, y_i) such that $x_i \in X$ and $y_i \in Y$, where the parameter ***marker='D'*** shows the points in the plot as a diamond.

```
>>> import matplotlib.pyplot
>>> matplotlib.pyplot.scatter([1,3,2],[5,4,6],color='gray')
<matplotlib.collections.PathCollection object at 0x120391c18>
>>> matplotlib.pyplot.show()
>>>
```

The example generates the Figure A.7.

matplotlib.pyplot.show()
shows the prepared graphics. The function **show()** caused the Figures A.2, A.3, A.6, and A.7 to appear.

matplotlib.pyplot.subplot(grid)
prepares a figure with a *grid* of subplots.

```
>>> import matplotlib.pyplot
>>> matplotlib.pyplot.subplots(2,3)
```

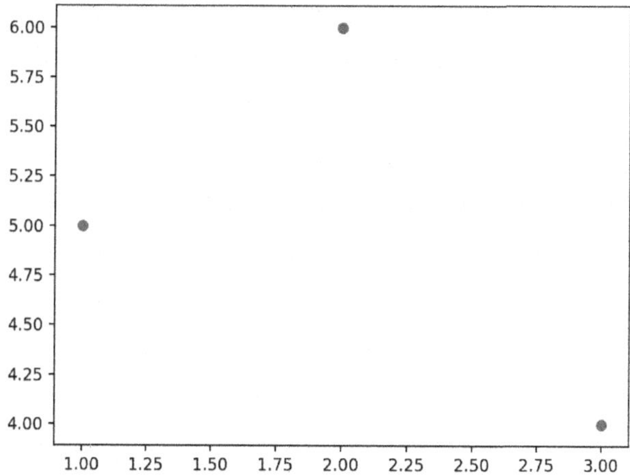

Figure A.7
matplotlib.pyplot.scatter

```
(<matplotlib.figure.Figure object at 0x10f962310>,
array([[<matplotlib.axes.AxesSubplot object at 0x115eae290>,
<matplotlib.axes.AxesSubplot object at 0x1174cb710>,
<matplotlib.axes.AxesSubplot object at 0x117548a50>],
[<matplotlib.axes.AxesSubplot object at 0x1175aca10>,
<matplotlib.axes.AxesSubplot object at 0x11762e910>,
<matplotlib.axes.AxesSubplot object at 0x117562910>]],
dtype=object))
>>> matplotlib.pyplot.show()
>>>
```

The example generates the Figure A.8.

matplotlib.pyplot.xlim(*left, right*)
sets the *left* limit and the *right* limit of the x-axis.

```
>>> import matplotlib.pyplot
>>> matplotlib.pyplot.xlim(1,5)
(1, 5)
>>> matplotlib.pyplot.show()
>>>
```

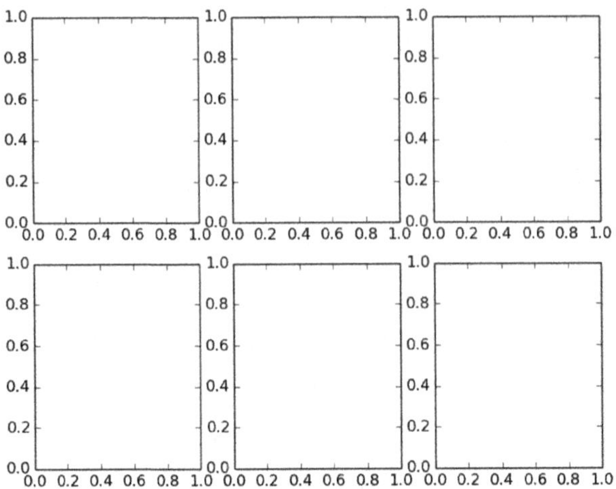

Figure A.8
matplotlib.pyplot.subplot

matplotlib.pyplot.xticks(*ticks, labels*, rotation=*r*)
sets the ticks of the x-axis in the position *ticks* with *labels* in a rotation angle *r*.

```
>>> import matplotlib.pyplot
>>> matplotlib.pyplot.xticks([1,2,3],['a','b','c'],rotation=90)
([<matplotlib.axis.XTick object at 0x114aa4410>,
<matplotlib.axis.XTick object at 0x11292f490>,
<matplotlib.axis.XTick object at 0x11304c3d0>],
<a list of 3 Text xticklabel objects>)
>>> matplotlib.pyplot.show()
>>>
```

Similar to `yticks`.

matplotlib.pyplot.ylim(*left, right*)
sets the *left* limit and the *right* limit of the y-axis.

```
>>> import matplotlib.pyplot
>>> matplotlib.pyplot.ylim(1,3)
(1, 3)
>>> matplotlib.pyplot.show()
>>>
```

neupy.algorithms.DiscreteBAM()
creates a discrete BAM network. Some methods are:

train(*X, y*) returns the neural network model trained with the sample *X* and the target values *y*.

predict(*X*) the same as the method **predict_output**.

predict_input(*y*) recovers the input *X* based on the output *y* and returns the input and output data.

predict_output(*X*) recovers the output *y* based on the input *X* and returns the input and output data.

```
>>> import neupy.algorithms
>>> import numpy
>>> a = numpy.matrix([0, 0])
>>> b = numpy.matrix([0, 1])
>>> c = numpy.matrix([1, 0])
>>> d = numpy.matrix([1, 1])
>>> data=numpy.concatenate([a,b,c,d])
>>> a = numpy.matrix([0])
>>> b = numpy.matrix([1])
>>> c = numpy.matrix([1])
>>> d = numpy.matrix([0])
>>> target=numpy.concatenate([a,b,c,d])
>>> NN = neupy.algorithms.DiscreteBAM()
>>> NN.train(data,target)
>>> q = numpy.matrix([1, 1])
>>> print('predictions:\n', NN.predict(q))
predictions:
(array([[1, 1]]), array([[0]]))
```

neupy.algorithms.DiscreteHopfieldNetwork()
returns a discrete Hopfield network. It has two important methods:

train(*array*) generates a neural network trained with *array*.

predict(*D*) returns the prediction based on the data *D*.

```
>>> import neupy.algorithms
>>> import numpy
>>> a = numpy.matrix([
... 0, 0, 0,
... 0, 1, 1,
... 1, 0, 1,
```

```
... 1, 1, 0])
>>> NN = neupy.algorithms.DiscreteHopfieldNetwork()
>>> NN.train(a)
>>> b = numpy.matrix([
... 0, 0, 0,
... 0, 1, 0,
... 1, 0, 0,
... 1, 1, 0])
>>> print('predictions:\n', NN.predict(b).reshape(4,3))
predictions:
[[0 0 0]
[0 1 1]
[1 0 1]
[1 1 0]]
>>>
```

neupy.algorithms.GRNN(std=*s*)
returns a generalized regression neural network (GRNN), where *s* is the standard deviation for the probability density function (PDF). Its methods are similar to **neupy.algorithms.PNN**.

```
>>> import neupy.algorithms
>>> x=    [[0, 0],
...        [0, 1],
...        [1, 0],
...        [1, 1]]
>>> y = [0, 1, 1, 0]
>>> NN = neupy.algorithms.GRNN()
>>> NN.train(x, y)
>>> print('predictions:', NN.predict(x))
predictions: [[7.44015195e-44]
 [1.00000000e+00]
 [1.00000000e+00]
 [7.44015195e-44]]
>>>
```

neupy.algorithms.PNN(std=*s*)
returns a probabilistic neural network (PNN), where *s* is the standard deviation for the probability density function (PDF). It has the methods:

train(*X*, *y*) returns the PNN model trained with the sample *X* and the target values *y*.

predict(*D*) returns the prediction of the PNN based on the date *D*.

```
>>> import neupy.algorithms
>>> x=     [[0, 0],
...          [0, 1],
...          [1, 0],
...          [1, 1]]
>>> y = [0, 1, 1, 0]
>>> NN = neupy.algorithms.PNN()
>>> NN.train(x, y)
>>> print('predictions:', NN.predict(x))
predictions: [0 1 1 0]
```

neupy.algorithms.SOFM(*parameters*)
creates a self-organizing map and returns it, where the *parameters* can be:

n_inputs = ***i*** defines the number *i* of input neurons.

learning_radius = ***r*** defines the neighborhood radius *r* to the winning neuron.

features_grid = ***g*** defines an n-dimensional grid *g*.

reduce_radius_after = ***a*** defines the number of epochs *a* to reduce the **learning_radius** by 1 until 0.

shuffle_data = ***L*** scrambles the data by default and when *L* is *True*, for not scrambling, use *False*.

verbose = **True** shows the training processing steps.

It has the methods:

train(*data*, epochs = ***i*)** returns the network trained with the *data* during *i* epochs.

predict(*data*) returns the prediction of the network based on the *data*.

```
>>> import neupy.algorithms
>>> import numpy
>>>
>>> data = numpy.array([
...     [-0, 0.2, 0.3],
...     [0.1, 0.8, 0.9],
...     [0.85, -0.3, 0.78],
...     [0.97, 0.85, 0.8],
... ])
>>>
```

```
>>> sofm = neupy.algorithms.SOFM(
...     n_inputs=3,
...     n_outputs=3,
... )
>>> sofm.train(data)
>>> print(sofm.predict(data))
[[0 1 0]
[0 0 1]
[0 0 1]
[0 0 1]]
>>>
```

numpy.arange(*start, stop, step*)

returns an array with $\frac{stop-start}{step}$ elements from *start* to *stop* and interval *step*.

```
>>> import numpy
>>> numpy.arange(1,6,2)
array([1, 3, 5])
>>>
```

numpy.argmax(*array, axis = a*)

returns the index of the largest element in an *array*, starting from zero. To search in a specific dimension, we just set *a*.

```
>>> import numpy
>>> numpy.argmax([[1,2],[3,4],[5,6],[7,8]])
7
>>> numpy.argmax([[1,2],[3,4],5])
1
>>> numpy.argmax([[1,2],[3,4],[5,6],[7,8]],axis=0)
array([3, 3])
>>> numpy.argmax([[1,2],[3,4],[5,6],[7,8]],axis=1)
array([1, 1, 1, 1])
```

numpy.concatenate(*[arrays], axis=a*)

returns a array that is the concatenation of the *arrays*, where *axis* defines how the *arrays* will be concatenated and *a* is the dimension, starting from the default zero, in which the elements are joined. If *a* is *None*, instead of an integer, the arrays will be flattened. Note that all other dimensions excluding axis *a* must match.

```
>>> import numpy
>>> a=numpy.array([[1,2],[3,4],[4,5]])
>>> a
array([[1, 2],
[3, 4],
[4, 5]])
>>> numpy.concatenate(a)
array([1, 2, 3, 4, 4, 5])
>>>
```

numpy.frombuffer(*buffer*)

returns a 1-dimensional array from a *buffer*.

```
>>> import numpy
>>> b=numpy.array([1,2,3])
>>> b
array([1, 2, 3])
>>> numpy.frombuffer(b)
array([  4.94065646e-324,   9.88131292e-324,   1.48219694e-323])
>>> numpy.frombuffer(b, dtype=int)
array([1, 2, 3])
>>>
```

numpy.linspace(*start, stop, n*)

returns an array with *n* elements evenly spaced from *start* to stop.

```
>>> import numpy
>>> numpy.linspace(1,5,3)
array([ 1.,  3.,  5.])
>>>
```

numpy.logical_xor(*X,Y*)

returns the truth value of the exclusive OR, either *X* or *Y* in terms of elements.

```
>>> import numpy
>>> x=numpy.array([1,2,3])
>>> y=numpy.array([1,0,4])
>>> numpy.logical_xor(x,y)
array([False,  True, False], dtype=bool)
>>> numpy.logical_xor(True,False)
```

```
True
>>> numpy.logical_xor(1,0)
True
>>> numpy.logical_xor(True,True)
False
>>> numpy.logical_xor(1,1)
False
>>>
```

numpy.matrix(*elements*)
returns a matrix of the elements.

```
>>> import numpy
>>> numpy.matrix([[1,2,3],[4,5,6]])
matrix([[1, 2, 3],
[4, 5, 6]])
>>> a=numpy.matrix([1,2,3])
>>> b=numpy.matrix([4,5,6])
>>> a*b.transpose()
matrix([[32]])
>>> a.transpose()*b
matrix([[ 4,  5,  6],
[ 8, 10, 12],
[12, 15, 18]])
>>>
```

numpy.meshgrid(*x, y*)
returns two matrices X and Y such that (X_{ij}, Y_{ij}) is the element (x_i, y_j) of the Cartesian product between x and y. Each row of X is given by x and each column of y is given by y.

```
>>> import numpy
>>> x=numpy.array([1,2,3])
>>> y=numpy.array([4,5])
>>> a,b = numpy.meshgrid(x,y)
>>> a
array([[1, 2, 3],
[1, 2, 3]])
>>> b
array([[4, 4, 4],
```

```
[5, 5, 5]])
>>>
```

numpy.ravel(*A*)

returns a 1-dimensional array of *A*.

```
>>> import numpy
>>> a=numpy.array([[1,2,3],[4,5,6]])
>>> a
array([[1, 2, 3],
[4, 5, 6]])
>>> numpy.ravel(a)
array([1, 2, 3, 4, 5, 6])
>>> a.ravel()
array([1, 2, 3, 4, 5, 6])
>>> numpy.array([[1,2,3],[4,5,6]])
array([[1, 2, 3],
[4, 5, 6]])
>>> numpy.array([[1,2,3],[4,5,6]]).ravel()
array([1, 2, 3, 4, 5, 6])
>>>
```

numpy.reshape(*A*, *r*, *c*)

returns an array with *r* rows and *c* columns based in the array *A*. We can also transform an array *A* with the syntax ***A*.reshape(*r*, *c*)**. The command ***A*.reshape(-1)** transforms *A* into a 1-dimensional array. We can also set -1 to either *r* or *c*, in case of either single feature or single sample.

```
>>> A = numpy.array([[1,2,3],[4,5,6]])
>>> numpy.reshape(A,-1)
array([1, 2, 3, 4, 5, 6])
>>> numpy.reshape(A,-1).reshape(3,2)
array([[1, 2],
[3, 4],
[5, 6]])
>>> numpy.reshape(A,-1).reshape(2,3)
array([[1, 2, 3],
[4, 5, 6]])
>>>
```

numpy.sin(*x*)

returns the $\sin(x)$ for a number x, but if x is an array, it returns an array with the sine of each element.

```
>>> import numpy
>>> numpy.sin(numpy.pi/2)
1.0
>>> numpy.sin([0,1,2,3])
array([ 0.     , 0.84147098, 0.90929743, 0.14112001])
>>>
```

numpy.transpose(*a*)

returns the transposition of an array *a*.

```
>>> import numpy
>>> numpy.array([[1,2,3],[4,5,6]])
array([[1, 2, 3],
[4, 5, 6]])
>>> numpy.array([[1,2,3],[4,5,6]]).transpose()
array([[1, 4],
[2, 5],
[3, 6]])
>>> a=numpy.array([[1,2,3],[4,5,6]])
>>> a
array([[1, 2, 3],
[4, 5, 6]])
>>> numpy.transpose(a)
array([[1, 4],
[2, 5],
[3, 6]])
>>> a.transpose()
array([[1, 4],
[2, 5],
[3, 6]])
>>>
```

numpy.zeros(*z*)

returns an array with *z* zeros.

```
>>> import numpy
>>> numpy.zeros(3)
array([ 0.,  0.,  0.])
>>>
```

pandas.DataFrame.head.head(*n*)

returns the first *n* elements of an object DataFrame. If *n* is not specified, the default is 5.

```
>>> import pandas
>>> f=pandas.read_csv("/Downloads/stocks.csv")
>>> f.head()
Date  Close
0  4/25/2011  15.77
1  4/26/2011  15.62
2  4/27/2011  15.35
3  4/28/2011  14.62
4  4/29/2011  14.75
>>> f.head(2)
Date  Close
0  4/25/2011  15.77
1  4/26/2011  15.62
>>>
```

pandas.Timestamp(*O*)

returns the object *O* in the format date and time for pandas object.

```
>>> import pandas
>>> pandas.Timestamp('2017.01.31')
Timestamp('2017-01-31 00:00:00')
>>>
```

pandas.read_csv("*F*")

reads a file *F* in CSV and returns its DataFrame.

```
>>> import pandas
>>> pandas.read_csv("/Downloads/stocks.csv")
Date      Close
```

```
0       4/25/2011  15.770000
1       4/26/2011  15.620000
2       4/27/2011  15.350000
3       4/28/2011  14.620000
4       4/29/2011  14.750000
5        5/2/2011  15.990000
6        5/3/2011  16.700001
7        5/4/2011  17.080000
8        5/5/2011  18.200001
9        5/6/2011  18.400000
10       5/9/2011  17.160000
11      5/10/2011  15.910000
12      5/11/2011  16.950001
13      5/12/2011  16.030001
14      5/13/2011  17.070000
15      5/16/2011  18.240000
16      5/17/2011  17.549999
17      5/18/2011  16.230000
18      5/19/2011  15.520000
19      5/20/2011  17.430000
20      5/23/2011  18.270000
21      5/24/2011  17.820000
22      5/25/2011  17.070000
23      5/26/2011  16.090000
24      5/27/2011  15.980000
25      5/31/2011  15.450000
26       6/1/2011  18.299999
27       6/2/2011  18.090000
28       6/3/2011  17.950001
29       6/6/2011  18.490000
...           ...        ...
1980     3/8/2019  16.049999
1981    3/11/2019  14.330000
1982    3/12/2019  13.770000
1983    3/13/2019  13.410000
1984    3/14/2019  13.500000
1985    3/15/2019  12.880000
1986    3/18/2019  13.100000
1987    3/19/2019  13.560000
1988    3/20/2019  13.910000
1989    3/21/2019  13.630000
1990    3/22/2019  16.480000
1991    3/25/2019  16.330000
1992    3/26/2019  14.680000
1993    3/27/2019  15.150000
1994    3/28/2019  14.430000
1995    3/29/2019  13.710000
1996     4/1/2019  13.400000
1997     4/2/2019  13.360000
1998     4/3/2019  13.740000
1999     4/4/2019  13.580000
2000     4/5/2019  12.820000
2001     4/8/2019  13.180000
2002     4/9/2019  14.280000
2003    4/10/2019  13.300000
```

```
2004  4/11/2019  13.020000
2005  4/12/2019  12.010000
2006  4/15/2019  12.320000
2007  4/16/2019  12.180000
2008  4/17/2019  12.600000
2009  4/18/2019  12.090000

[2010 rows x 2 columns]
>>>
```

pandas.set_index(*i*)

sets *i* as the index of the DataFrame.

```
>>> import pandas
>>> f=pandas.read_csv("/Downloads/stocks.csv")
>>> f.set_index(['Date'])
Close
Date
4/25/2011  15.770000
4/26/2011  15.620000
4/27/2011  15.350000
4/28/2011  14.620000
4/29/2011  14.750000
5/2/2011   15.990000
5/3/2011   16.700001
5/4/2011   17.080000
5/5/2011   18.200001
5/6/2011   18.400000
5/9/2011   17.160000
5/10/2011  15.910000
5/11/2011  16.950001
5/12/2011  16.030001
5/13/2011  17.070000
5/16/2011  18.240000
5/17/2011  17.549999
5/18/2011  16.230000
5/19/2011  15.520000
5/20/2011  17.430000
5/23/2011  18.270000
5/24/2011  17.820000
5/25/2011  17.070000
5/26/2011  16.090000
5/27/2011  15.980000
5/31/2011  15.450000
6/1/2011   18.299999
6/2/2011   18.090000
6/3/2011   17.950001
6/6/2011   18.490000
...              ...
3/8/2019   16.049999
3/11/2019  14.330000
3/12/2019  13.770000
```

```
3/13/2019  13.410000
3/14/2019  13.500000
3/15/2019  12.880000
3/18/2019  13.100000
3/19/2019  13.560000
3/20/2019  13.910000
3/21/2019  13.630000
3/22/2019  16.480000
3/25/2019  16.330000
3/26/2019  14.680000
3/27/2019  15.150000
3/28/2019  14.430000
3/29/2019  13.710000
4/1/2019   13.400000
4/2/2019   13.360000
4/3/2019   13.740000
4/4/2019   13.580000
4/5/2019   12.820000
4/8/2019   13.180000
4/9/2019   14.280000
4/10/2019  13.300000
4/11/2019  13.020000
4/12/2019  12.010000
4/15/2019  12.320000
4/16/2019  12.180000
4/17/2019  12.600000
4/18/2019  12.090000

[2010 rows x 1 columns]
>>>
```

pandas.to_datetime(*O*)

returns the object *O* in the format Python date and time object.

```
>>> import pandas
>>> f=pandas.read_csv("/Downloads/stocks.csv")
>>> pandas.to_datetime(f['Date'])
0       2011-04-25
1       2011-04-26
2       2011-04-27
3       2011-04-28
4       2011-04-29
5       2011-05-02
6       2011-05-03
7       2011-05-04
8       2011-05-05
9       2011-05-06
10      2011-05-09
11      2011-05-10
12      2011-05-11
13      2011-05-12
14      2011-05-13
```

```
15      2011-05-16
16      2011-05-17
17      2011-05-18
18      2011-05-19
19      2011-05-20
20      2011-05-23
21      2011-05-24
22      2011-05-25
23      2011-05-26
24      2011-05-27
25      2011-05-31
26      2011-06-01
27      2011-06-02
28      2011-06-03
29      2011-06-06
...
1980    2019-03-08
1981    2019-03-11
1982    2019-03-12
1983    2019-03-13
1984    2019-03-14
1985    2019-03-15
1986    2019-03-18
1987    2019-03-19
1988    2019-03-20
1989    2019-03-21
1990    2019-03-22
1991    2019-03-25
1992    2019-03-26
1993    2019-03-27
1994    2019-03-28
1995    2019-03-29
1996    2019-04-01
1997    2019-04-02
1998    2019-04-03
1999    2019-04-04
2000    2019-04-05
2001    2019-04-08
2002    2019-04-09
2003    2019-04-10
2004    2019-04-11
2005    2019-04-12
2006    2019-04-15
2007    2019-04-16
2008    2019-04-17
2009    2019-04-18
Name: Date, Length: 2010, dtype: datetime64[ns]
>>>
```

scipy.special.comb(*k*, *n*, *exact=True*))

returns the combination of *n* choose *k*. The parameter *exact=True* returns a long integer instead of rounding with floating-point precision.

```
>>> import scipy.special
>>> scipy.special.comb(5, 2)
array(9.999999999999998)
>>> scipy.special.comb(5, 2, exact=True)
10
>>>
```

scipy.stats.norm.pdf(*X*)

return an array with the probability density function (PDF) of each element of the array *X*, i.e., for each $x \in X$, the function computes

$$\frac{\exp\left(-\frac{x^2}{2}\right)}{\sqrt{2\pi}}.$$

```
>>> import scipy.stats
>>> scipy.stats.norm.pdf([1,3,5])
array([ 2.41970725e-01, 4.43184841e-03, 1.48671951e-06])
>>>
```

scipy.stats.norm.rvs(*loc=l, scale=s, size=n*)

generates random variables for a continuous statistical distribution, where the location *loc* represents the mean *l*, the *scale* represents the standard deviation *s*, and the *size* determines the number *n* of variables returned in the a array.

```
>>> import scipy.stats
>>> scipy.stats.norm.rvs(loc=9, scale=.1, size=4)
array([ 9.06958065, 8.95592511, 9.07394299, 8.90148694])
>>> scipy.stats.norm.rvs(loc=9, scale=.1, size=4)
array([ 8.99935205, 9.12326591, 9.00818677, 9.05582916])
>>>
```

sklearn.datasets.fetch_openml(name = "*D*")

downloads and returns the dataset *D* from the repository OpenML[1].

```
>>> import sklearn.datasets
>>> D=sklearn.datasets.fetch_openml('tic-tac-toe')
>>> D
```

[1] https://openml.org/

sklearn.datasets.load_digits()
loads and returns a dataset with 8×8 images of a digit for classification.

```
>>> import sklearn.datasets
>>> digits = sklearn.datasets.load_digits()
>>> print(digits.data)
[[ 0.   0.   5. ...   0.   0.   0.]
[ 0.   0.   0. ... 10.   0.   0.]
[ 0.   0.   0. ... 16.   9.   0.]
...
[ 0.   0.   1. ...   6.   0.   0.]
[ 0.   0.   2. ... 12.   0.   0.]
[ 0.   0. 10. ... 12.   1.   0.]]
>>> print(digits.data.shape)
(1797, 64)
>>>
```

sklearn.datasets.make_moons(*n_samples=s*)
returns an array of samples X with $2s$ elements and an array of labels y with s elements. If the parameter *n_samples* is not defined, the command generates by default 100 points. The parameter *noise=n* adds the standard deviation of Gaussian noise to the data X. If we want to avoid shuffling data X, we need to use the parameter *shuffle=False*. If we want to reproduce the experiment, we need to set the random state with an integer, e.g., using the parameter *random_state=0*.

```
>>> import sklearn.datasets
>>> X,y=sklearn.datasets.make_moons(n_samples=10)
>>> X
array([[ 6.12323400e-17,  1.00000000e+00],
[ 1.70710678e+00, -2.07106781e-01],
[ 0.00000000e+00,  5.00000000e-01],
[ 2.92893219e-01, -2.07106781e-01],
[ 2.00000000e+00,  5.00000000e-01],
[ 7.07106781e-01,  7.07106781e-01],
[ 1.00000000e+00,  0.00000000e+00],
[-7.07106781e-01,  7.07106781e-01],
[-1.00000000e+00,  1.22464680e-16],
[ 1.00000000e+00, -5.00000000e-01]])
>>> y
array([0, 1, 1, 1, 1, 0, 0, 0, 0, 1])
>>>
```

sklearn.decomposition.PCA(n_components = *c*, random_state = *0*)
returns a PCA with *c* components. We use random_state = *0* to achieve a reproducible experiment. The method **fit_transform(*D*)** applies the PCA in the data *D*.

```
>>> import sklearn.decomposition
>>> import numpy
>>> a=numpy.array([[1,2,3],[4,5,6],[7,8,9]])
>>> PCA=sklearn.decomposition.PCA(n_components=2)
>>> PCA.fit_transform(a)
array([[ 5.19615242e+00,  6.24556048e-17],
[-0.00000000e+00, -0.00000000e+00],
[-5.19615242e+00,  6.24556048e-17]])
>>> a
array([[1, 2, 3],
[4, 5, 6],
[7, 8, 9]])
>>>
```

sklearn.model_selection.train_test_split(*A*, shuffle=False)
splits array *A* into four arrays and returns arrays of sample and labels for training and test. We use **shuffle=False** to get the same result.

```
>>> import sklearn.model_selection
>>> a=range(10)
>>> sklearn.model_selection.train_test_split(a)
[[3, 5, 1, 0, 4, 8, 2], [7, 6, 9]]
>>> sklearn.model_selection.train_test_split(a,shuffle=False)
[[0, 1, 2, 3, 4, 5, 6], [7, 8, 9]]
>>>
```

sklearn.neural_network.MLPClassifier(*activation='f', solver='s'*)
returns an artificial neural network based on multi-layer perceptron classifier. The activation function *f* can be:

identity – the linear bottleneck, i.e., $f(x) = x$;

logistic – the logistic sigmoid function, i.e., $f(x) = 1/(1 + \exp(-x))$;

tanh – the hyperbolic tangent function, i.e., $f(x) = \tanh(x)$;

relu – the rectified linear unit function, i.e., $f(x) = \max(0, x)$.

If the parameter activation is not declared, the default function is **relu**.

The solver *s* for weight optimization can be:

lbfgs – the Broyden-Fletcher-Goldfarb-Shanno algorithm, a quasi-Newton method;

sgd – the stochastic gradient descent;

adam – a stochastic gradient-based optimizer.

We can use other parameters, for instance, **hidden_layer_sizes** whose the default is 100 neurons in one hidden layer. The parameter *hidden_layer_sizes=(500, 400, 300, 200, 100)* sets five hidden layers containing from 500 to 100 neurons, respectively. The parameter **verbose** = ***True*** shows the training processing steps.

This command has four important methods:

fit(*X*, *y*) returns the neural network model trained with the sample *X* and the target values *y*.

predict(*D*) returns the predicted values based on the input data *D*.

predict_proba(*D*) returns the probability of the predicted values based on the input data *D*.

score(*X*,*y*) return the accuracy on X and y.

```
>>> import sklearn.neural_network
>>> x = [[0, 0],
...      [0, 1],
...      [1, 0],
...      [1, 1]]
>>> y = [0, 1, 1, 0]
>>> NN = sklearn.neural_network.MLPClassifier(activation='tanh',
...      solver='lbfgs')
>>> NN.fit(x, y)
MLPClassifier(activation='tanh', alpha=0.0001, batch_size='auto',
beta_1=0.9,beta_2=0.999,early_stopping=False,epsilon=1e-08,
hidden_layer_sizes=(100,), learning_rate='constant',
learning_rate_init=0.001, max_iter=200, momentum=0.9,
n_iter_no_change=10, nesterovs_momentum=True, power_t=0.5,
random_state=None, shuffle=True,solver='lbfgs',tol=0.0001,
validation_fraction=0.1, verbose=False, warm_start=False)
>>> print('score:', NN.score(x, y))
score: 1.0
>>> print('predictions:', NN.predict(x))
predictions: [0 1 1 0]
>>>
```

sklearn.neural_network.MLPRegressor()

returns an artificial neural network based on multi-layer perceptron regressor. With exception of **predict_proba(*D*)**, the parameters and methods are similar to the command **sklearn.neural_network.MLPClassifier()**. In this case, the method **score(*X*,*y*)** returns the R-squared value, instead of accuracy.

```
>>> import sklearn.neural_network
>>> x = [[0, 0],
...      [0, 1],
...      [1, 0],
...      [1, 1]]
>>> y = [0, 1, 1, 0]
>>> NN = sklearn.neural_network.MLPRegressor(activation='tanh',
...      solver='lbfgs')
>>> NN.fit(x, y)
MLPRegressor(activation='tanh', alpha=0.0001, batch_size='auto',
beta_1=0.9, beta_2=0.999,early_stopping=False,epsilon=1e-08,
hidden_layer_sizes=(100,), learning_rate='constant',
learning_rate_init=0.001, max_iter=200, momentum=0.9,
n_iter_no_change=10, nesterovs_momentum=True, power_t=0.5,
random_state=None, shuffle=True, solver='lbfgs', tol=0.0001,
validation_fraction=0.1, verbose=False, warm_start=False)
>>> print('score:', NN.score(x, y))
score: 0.9999988676278221
>>> print('predictions: \n', NN.predict(x))
predictions:
[3.50179813e-04 9.99600703e-01 9.99358686e-01 6.62589540e-04]
>>>
```

sys.argv[*n*]

returns the string of the command line argument number *n*. In the following example, the system does not have parameter because it was run in the command line.

```
>>> import sys
>>> sys.argv[0]
''
>>>
```

time.process_time()

returns processing time of the current process in a float of seconds.

```
>>> import time
>>> time.process_time()
1.244904
>>> time.process_time()-time.process_time()
-6.000000000172534e-06
>>>
```

Bibliography

[1] Abdi, H., Edelman, H., Valentin, D., and Edelman, B. (1999). *Neural Networks*. Number 124 in Quantitative Applications in Social Science Series. Publications.

[2] Abramovich, F., Barley, T. C., and Sapatinas, T. (2000). Wavelet analysis and its statistical application. *Journal of the Royal Statistical Society D*, 49(1):1–29.

[3] Adams, J. B. (1999). Predicting pickle harvests using parametric feedfoward network. *Journal of Applied Statistics*, 26:165–176.

[4] Addeh, A., Khormali, A., and Golilarz, N. A. (2018). Control chart pattern recognition using RBF neural network with new training algorithm and practical features. *ISA Transactions*, 79:202–216.

[5] Addrians, P. and Zanringe, D. (1996). *Data Mining*. Addison-Wesley, 2nd edition.

[6] Aitkin, M. and Foxall, R. (2003). Statistical modelling of artificial neural networks using the multilayer perceptron. *Statistics and Computing*, 13:227–239.

[7] Amari, S. (1993). Network and chaos: Statistical and probabilistic aspects. In *Mathematical Methods of Neurocomputing*, pages 1–39. Chapman & Hall,.

[8] Arminger, G. and Enache, D. (1996). Statistical models and artificial neural networks. In Bock, H.-H. and Polasek, W., editors, *Data Analysis and Information Systems*, pages 243–260, Heidelberg. Springer.

[9] Arminger, G., Enache, D., and Bonne, T. (1997). Analyzing credit risk data: a comparision of logistic discrimination, classification, tree analysis and feedfoward networks. *Computational Statistics*, 12:293–310.

[10] Arminger, G. and Polozik, W. (1996). Data analysis and information systems, statistical and conceptual approaches. studies in classification data analysis and knowledge organization. In *Statistical Models and Neural Networks*, volume 7, pages 243–260. Springer.

[11] Asparoukhov, O. K. and Krzanowski, W. J. (2001). A comparison of discriminant procedures for binary variables. *Computational Statistics and Data Analysis*, 38:139–160.

[12] Bakker, B., Heskes, T., Neijt, J., and Kappen, B. (2004). Improving Cox survival analysis with a neural-Bayesian approach. *Statistics in Medicine*, 23(19):2989–3012.

[13] Balakrishnan, P. V. S., Cooper, M. C., Jacob, M., and Lewis, P. A. (1994). A study of the classification capabilities of neural networks using unsupervised learning: a comparision with k-means clustering. *Psychometrika*, 59(4):509–525.

[14] Balkin, S. D. and Ord, J. K. (2000). Automatic neural network modeling for univariate time series. *International Journal of Forecasting*, 16:509–515.

[15] Balkin, S. D. J. and Lim, D. K. J. (2000). A neural network approach to response surface methodology. *Communications in Statistics - Theory and Methods*, 29(9/10):2215–2227.

[16] Belue, L. M. and Bauer, K. W. (1999). Designing experiments for single output multilayer perceptrons. *Journal of Statistical Computation and Simulation*, 64:301–332.

[17] Ben-David, S., Hrubeš, P., Moran, S., Shpilka, A., and Yehudayoff, A. (2019). Learnability can be undecidable. *Nature Machine Intelligence*, 1(1):44.

[18] Biganzoli, E., Baracchi, P., and Marubini, E. (1998). Feed forward neural networks for the analysis of censored survival data: A partial logistic regression approach. *Statistics in Medicine*, 17:1169–1186.

[19] Biganzoli, E., Baracchi, P., and Marubini, E. (2002). A general framework for neural network models on censored survival data. *Neural Networks*, 15:209–218.

[20] Bishop, C. M. (1995). *Neural Network for Pattern Recognition*. Clarendon Press, Oxford.

[21] Blake, A. P. and Kapetanios, G. (2003). A radial basis function artificial neural network test for neglected nonlinearity. *The Econometrics Journal*, 6(2):357–373.

[22] Boll, D., Geomini, P. A. M. J., Brölmann, H. A. M., Sijmons, E. A., Heintz, P. M., and Mol, B. W. J. (2003). The pre-operative assessment of the adnexal mass: The accuracy of clinical estimates versus clinical prediction rules. *International Journal of Obstetrics and Gynecology*, 110:519–523.

[23] Borges de Oliveira, F. (2017). *On Privacy-Preserving Protocols for Smart Metering Systems: Security and Privacy in Smart Grids*. International Publishing, Cham.

[24] Bose, N. K. and Liang, P. (1996). *Neural Networks Fundamentals with Graphs, Algorithms and Applications*. McGraw-Hill Inc., 2nd edition.

[25] Bose, S. (2003). Multilayer statistical classifiers. *Computational Statistics and Data Analysis*, 42:685–701.

[26] Bounds, D. G., Lloyd, P. J., and Mathew, B. G. (1990). A comparision of neural network and other pattern recognition aproaches to the diagnosis of low back disorders. *Neural Networks*, 3:583–591.

[27] Braga, A. P., Ludemir, T. B., and Carvalho, A. C. P. L. F. (2000). *Redes Neurais Artificiais: Teoria e Aplicações*. Editora LTC, São Paulo.

[28] Brockett, P. H., Cooper, W. W., Golden, L. L., and Xia, X. (1997). A case study in applying neural networks to predicting insolvency for property and causality insurers. *Journal of the Operational Research Society*, 48:1153–1162.

[29] Calôba, G. M., Calôba, L. P., and Saliby, E. (2002). Cooperação entre redes neurais artificiais e técnicas 'clássicas' para previsão de demanda de uma série de vendas de cerveja na austrália. *Pesquisa Operacional*, 22(3):345–358.

[30] Calvo, R. (1997). Factor analysis in educational research: An artificial neural network perspective. In *Proceedings of PACES/SPICIS 97*.

[31] Carvalho, L. V. (1997). *Data Mining*. Editora Erika, São Paulo.

[32] Carvalho, M. C. M., Dougherty, M. S., Fowkes, A. S., and Wardman, M. R. (1998). Forecasting travel demand: A comparison of logit and artificial neural network methods. *Journal of the Operational Research Society*, 49:717–722.

[33] Cha, S. M. and Chan, L. E. (2002). Applying independent component analysis to factor model in finance. In *Intelligent Data Engineering and Automated Learning, IDEAL 2000, Data Mining, Finacncial Enginerring and Inteligent Agents*, pages 538–544, Bruges. Springer.

[34] Chakraborthy, K., Mehrotra, K., Mohan, C. K., and Ranka, S. (1992). Forecasting the behavior of multivariate time series using neural networks. *Neural Networks*, 5:961–970.

[35] Chan, L. W. and Cha, S. M. (2001). Selection of independent factor model in finance. In *Proceedings of 3rd International Conference on Independent Component Analyses*, pages 161–166.

[36] Cheng, B. and Titterington, D. M. (1994). Neural networks: A review from a statistical perspective (with discussion). *Statistical Sciences*, 9:2–54.

[37] Cheng, C. S. (1997). A neural network approach for the analysis of control charts patterns. *International Journal of Production Research*, 35:667–697.

[38] Chester, M. (1993). *Neural Networks: A Tutorial*. Prentice Hall.

[39] Chiu, C. C., Yeh, S. J., and Chen, C. H. (2000). Self-organizing arterial pressure pulse classification using neural networks: theoretical considerations and clinical applicability. *Computers in Biology and Medicine*, 30(2):71–88.

[40] Chryssolouriz, G., Lee, M., and Ramsey, A. (1996). Confidence interval predictions for neural networks models. *IEEE Transactions on Neural Networks*, 7:229–232.

[41] Church, K. B. and Curram, S. P. (1996). Forecasting consumer's expenditure: A comparison between econometric and neural networks models. *International Journal of Forecasting*, 12:255–267.

[42] Ciampi, A. and Etezadi-Amoli, J. (1985). A general model for testing the proportional harzards and the accelerated failure time hypothesis in the analysis of censored survival data with covariates. *Communications in Statistics A*, 14:651–667.

[43] Ciampi, A. and Lechevalier, Y. (1997). Statistical models as building blocks of neural networks. *Communications in Statistics - Theory and Methods*, 26(4):991–1009.

[44] Ciampi, A. and Zhang, F. (2002). A new approach to training back-propagation artificial neural networks: Empirical evaluation on ten data sets from clinical studies. *Statistics in Medicine*, 21:1309–1330.

[45] Cigizoglu, H. K. (2003). Incorporation of ARIMA models into flow forecasting by artificial neural networks. *Environmetrics*, 14(4):417–427.

[46] Clausen, S. E. (1998). *Applied Correspondence Analysis: An Introduction*. Number 121 in Quantitative Applications in Social Science Series. SAGE Publications.

[47] Comon, P. (1994). Independent component analysis, a new concept? *Signal Processing*, 3:287–314.

[48] Cooper, J. C. B. (1999). Artificial neural networks versus multivariate statistics: An application from economics. *Journal of Applied Statistics*, 26:909–921.

[49] Copobianco, E. (2000). Neural networks and statistical inference: seeking robust and efficient learning. *Journal of Statistics and Data Analysis*, 32:443–445.

[50] Correia-Perpinan, M. A. (1997). A review of dimensional reduction techniques. Techinical report CS-96-09, University of Sheffield.

[51] Coutinho, M., de Oliveira Albuquerque, R., Borges, F., García Villalba, L. J., and Kim, T.-H. (2018). Learning perfectly secure cryptography to protect communications with adversarial neural cryptography. *Sensors*, 18(5).

[52] Cox, T. and Cox, M. (2000). *Multidimensional Scaling, Second Edition.* Chapman & Hall/CRC Monographs on Statistics & Applied Probability. CRC Press.

[53] Cramer, H. (1939). On the representation of functions by certain Fourier integrals. *Transactions of the American Math. Society*, 46:191–201.

[54] Darmois, G. (1953). Analyse générale des liaisons stochastiques: etude particulière de l'analyse factorielle linéaire. *Review of the International Statistical Institute*, 21(1/2):2–8.

[55] De Veaux, R. D., Schweinsberg, J., Schumi, J., and Ungar, L. H. (1998). Prediction intervals for neural networks via nonlinear regression. *Technometrics*, 40(4):273–282.

[56] Delichere, M. and Memmi, D. (2002a). Analyse factorielle neuronale pour documents textuels. In *Traiment Automatique des Languages Naturelles.*

[57] Delichere, M. and Memmi, D. (2002b). Neural dimensionality reduction for document processing. In *European Symposium on Artificial Neural Networks.*

[58] Desai, V. S., Crook, J. N., and Overstreet, G. A. (1996). A comparison of neural networks and linear scoring models in the credit union environment. *European Journal of Operational Research*, 95:24–37.

[59] Diamantaras, K. I. and Kung, S. Y. (1996). *Principal Components Neural Networks. Theory and Applications*. Wiley.

[60] Donaldson, R. G. and Kamstra, M. (1996). Forecasting combining with neural networks. *Journal of Forecasting*, 15:49–61.

[61] Draghici, S. and Potter, R. B. (2003). Predicting HIV drug resistance with neural networks. *Bioinformatics*, 19(1):98–107.

[62] Dunteman, G. H. (1989). *Principal Component Analysis*. Number 69 in Quantitative Applications in Social Science Series. SAGE Publications.

[63] Ennis, M., Hinton, G., Naylor, D., Revow, M., and Tibshirani, R. (1998). A comparison of statistical learning methods on the GUSTO database. *Statistics In Medicine*, 17:2501–2508.

[64] Faraggi, D., LeBlanc, M., and Crowley, J. (2001). Understanding neural networks using regression trees: An application to multiply myeloma survival data. *Statistics in Medicine*, 20:2965–2976.

[65] Faraggi, D. and Simon, R. (1995a). The maximum likelihood neural network as a statistical classification model. *Journal of Statistical Planning and Inference*, 46:93–104.

[66] Faraggi, D. and Simon, R. (1995b). A neural network model for survival data. *Statistics in Medicine*, 14:73–82.

[67] Faraggi, D., Simon, R., Yaskily, E., and Kramar, A. (1997). Bayesian neural network models for censored data. *Biometrical Journal*, 39:519–532.

[68] Faraway, J. and Chatfield, C. (1998). Time series forecasting with neural networks: A comparative study using the airline data. *Applied Statistics*, 47:231–250.

[69] Fausett, L. (1994). *Fundamentals of Neural Networks Architectures, Algorithms and Applications*. Prentice Hall.

[70] Finney, D. J. (1947). *Probit Analysis: A Statistical Treatment of the Sigmoid Response Curve*. Macmillan, Oxford.

[71] Fiori, S. (2003). Overview of independent component analysis technique with application to synthetic aperture radar (SAR) imagery processing. *Neural Networks*, 16:453–467.

[72] Fisher, R. A. (1936). The use of multiples measurements in taxonomic problems. *Annals of Eugenics*, 7:179–188.

[73] Fletcher, D. and Goss, E. (1993). Forecasting with neural networks: An application using bankruptcy data. *Information and Management*, 24:159–167.

[74] Fox, J. (2000a). *Multiple and Generalized Nonparametric Regression*, volume 131 of *Quantitative Applications in Social Science Series*. SAGE Publications.

[75] Fox, J. (2000b). *Nonparametric Simple Regression*, volume 130 of *Quantitative Applications in Social Science Series*. SAGE Publications.

[76] Ghiassi, M., Saidane, H., and Zimbra, D. K. (2005). A dynamic artificial neural network model for forecasting series events. *International Journal of Forecasting*, 21:341–362.

[77] Giannakopoulos, X., Karhunen, J., and Oja, E. (1999). An experimental comparision of neural algorithms for independent component analysis and blind separation. *International Journal of Neural Systems*, 9:99–114.

[78] Girolami, M. (1999). *Self-Organizing Neural Networks: Independent Component Analysis and Blind Source Separation*. Springer.

[79] Gnandesikan, R. (1991). *Methods of Statistical Data Analysis of Multivariate Observations*. Wiley, 2nd edition.

[80] Gorr, W. L., Nagin, D., and Szczypula, J. (1994). Comparative study of artificial neural networks and statistical models for predicting student grade point average. *International Journal of Forecasting*, 10:17–34.

[81] Groves, D. J., Smye, S. W., Kinsey, S. E., Richards, S. M., Chessells, J. M., Eden, O. B., and Basley, C. C. (1999). A comparison of Cox regression and neural networks for risk stratification in cases of acute lymphoblastic leukemia in children. *Neural Computing and Applications*, 8:257–264.

[82] Gurney, K. (1997). *An Introduction to Neural Networks*. Taylor & Francis, Inc., Abingdon.

[83] Hammad, T. A., Abdel-Wahab, M. F., DeClaris, N., El-Sahly, A., El-Kady, N., and Strickland, G. T. (1996). Comparative evaluation of the use of neural networks for modeling the epidemiology of Schistosomiasis mansoni. *Transactions of the Royal Society of Tropical Medicine and Hygiene*, 90:372–376.

[84] Hanke, M., Halchenko, Y. O., Sederberg, P. B., Hanson, S. J., Haxby, J. V., and Pollmann, S. (2009). PyMVPA: a Python toolbox for multivariate pattern analysis of fMRI data. *Neuroinformatics*, 7(1):37–53.

[85] Haykin, S. (1999). *Neural Networks: A Comprehensive Foundation*. Prentice Hall, 2nd edition.

[86] Haykin, S. S. (2009). *Neural Networks and Learning Machines*. Pearson Education, Upper Saddle River, third edition.

[87] Hertz, J., Krogh, A., and Palmer, R. G. (1991). *Introduction to the Theory of Neural Computation*. Addison-Wesley.

[88] Hill, T., Marquez, L., O'Connor, M., and Remus, W. (1994). Artificial neural network models for forecasting and decision making. *International Journal of Forecasting*, 10:5–15.

[89] Hochreiter, S. and Schmidhuber, J. (1997). Long short-term memory. *Neural Computation*, 9(8):1735–1780.

[90] Hsu, A. L., Tang, S. L., and Halgamuge, S. K. (2003). An unsupervised hierarchical dynamic self-organizing approach to cancer class discovery and marker gene identification in microarray data. *Bioinformatics*, 19:2131–2140.

[91] Hubbard, B. B. (1998). *The World According to Wavelets: The Story of a Mathematical Technique in the Making, Second Edition*. Taylor & Francis.

[92] Hwang, J. T. G. and Ding, A. A. (1997). Prediction interval for artificial neural networks. *Journal of the American Statistical Association*, 92:748–757.

[93] Hwarng, H. B. (2002). Detecting mean shift in AR(1) process. *Decision Sciences Institute 2002 Annual Meeting Proceedings*, pages 2395–2400.

[94] Hyvärinen, A. (1999). Survey on independent component analysis. *Neural Computing Surveys*, 2:94–128.

[95] Hyvärinen, A., Karhunen, J., and Oja, E. (2001). *Independent Component Analysis*. Wiley, New York.

[96] Hyvärinen, A. and Oja, E. (2000). Independent component analysis: algorithm and applications. *Neural Networks*, 13:411–430.

[97] Insua, D. R. and Müller, P. (1998). *Feedforward Neural Networks for Nonparametric Regression*, pages 181–193. Springer, New York.

[98] Ishihara, S., Ishihara, K., Nagamashi, M., and Matsubara, Y. (1995). An automatic builder for a Kansei engineering expert systems using a self-organizing neural networks. *International Journal of Industrial Ergonomics*, 15(1):13–24.

[99] Iyengar, S. S., Cho, E. C., and Phola, V. V. (2002). *Foundations of Wavelets Networks and Applications*. Chapman & Hall.

[100] Jackson, J. E. (1991). *A User's Guide to Principal Components*. Wiley.

[101] Jordan, M. I. (1995). Why the logistic function? a tutorial discussion on probabilities and neural networks. Report 9503, MIT Computational Cognitive Science.

[102] Jordan, M. I. and Jacobs, R. A. (1994). Hierarchical mixtures of experts and the EM algorithm. *Neural Computation*, 6:181–214.

[103] Jutten, C. and Taleb, A. (2000). Source separation: from dust till down. In *Procedings ICA 2000*.

[104] Kagan, A. M., Linnik, Y., and Rao, C. R. (1973). *Characterization Problems in Mathematical Statistics*. Wiley.

[105] Kajitani, Y., Hipel, K. W., and Mcleod, A. I. (2005). Forecasting nonlinear time series with feed-forward neural networks: a case study of Canadian lynx data. *Journal of Forecasting*, 24(2):105–117.

[106] Kalbfleish, J. D. and Prentice, R. L. (2002). *The Statistical Analysis of Failure Data*. Wiley, 2nd edition.

[107] Karhunen, J., Oja, E., Wang, L., and Joutsensalo, J. (1997). A class of neural networks for independent component analysis. *IEEE Transactions on Neural Networks*, 8:486–504.

[108] Kartalopulos, S. V. (1996). *Understanding Neural Networks and Fuzzy Logic. Basic Concepts and Applications*. IEEE Press.

[109] Kasabov, N. K. (1996). *Foundations of Neural Networks. Fuzzy Systems and Knowledge Engineering*. MIT Press.

[110] Kiani, K. (2003). On performance of neural networks nonlinearity tests: Evidence from G5 countries. *Department of Economics, Kansas State University*.

[111] Kim, S. H. and Chun, S. H. (1998). Graded forecasting using an array of bipolar predictions: application of probabilistic neural networks to a stock market index. *International Journal of Forecasting*, 14(3):323–337.

[112] Kin, J. O. and Mueller, C. W. (1978). *Introduction to Factor Analysis*. Number 13 in Quantitative Applications in Social Science Series. SAGE Publications.

[113] Kuan, C.-M. and White, H. (1994). Artificial neural networks: An econometric perspective. *Econometric Reviews*, 13:1–91.

[114] Lai, T. L. and Wong, S. P. S. (2001). Stochastic neural networks with applications to nonlinear time series. *Journal of the American Statistical Association*, 96:968–981.

[115] Laper, R. J. A., Dalton, K. J., Prager, R. W., Forsstrom, J. J., Selbmann, H. K., and Derom, R. (1995). Application of neural networks to the ranking of perinatal variables influencing birthweight. *Scandinavian Journal of Clinical and Laboratory Investigation*, 55 (suppl 22):83–93.

[116] Lapuerta, P., Azen, S. P., and LaBree, L. (1995). Use of neural networks in predicting the risk of coronary artery disease. *Computer and Biomedical Research*, 28:38–52.

[117] Lebart, L. (1997). Correspondence analysis and neural networks. In *Proceedings of Fourth International Meeting on Multidimensional Data Analysis*, pages 47–58. K. Fernandes Aquirre (Ed).

[118] Lee, H. K. H. (1998a). *Model Selection and Model Averaging for Neural Network*. Ph.D. thesis, Department of Statistics, Carnegie Mellon University.

[119] Lee, H. K. H. (2004). *Bayesian Nonparametrics via Neural Networks*. Society for Industrial and Applied Mathematics.

[120] Lee, T. H. (2001). Neural network test and nonparametric kernel test for neglected nonlinearity in regression models. *Studies in Nonlinear Dynamics and Econometrics*, 4.

[121] Lee, T. H., White, H., and Granger, C. W. J. (1993). Testing for neglected nonlinearity in time series models. *Journal of Econometrics*, 56:269–290.

[122] Lee, T. W. (1998b). *Independent Component Analysis Theory and Applications*. Kluwer Academic Publishers.

[123] Lee, T. W., Girolami, M., and Sejnowski, T. J. (1999). Independent component analysis using an extended informax algorithm for mixed subgaussian and supergaussian sources. *Neural Computation*, 11:417–441.

[124] Lenard, M. J., Alam, P., and Madey, G. R. (1995). The application of neural networks and a qualitative response model to the auditor's going concern uncertainty decision. *Decision Sciences*, 26:209–226.

[125] Li, X., Ang, C. L., and Gray, R. (1999). Stochastic neural networks with applications to nonlinear time series. *Journal of Forecasting*, 18:181–204.

[126] Liestol, K., Andersen, P. K., and Andersen, U. (1994). Survival analysis and neural nets. *Statistics in Medicine*, 13:1189–1200.

[127] Lippmann, R. (1988). Neutral nets for computing. In *ICASSP-88., International Conference on Acoustics, Speech, and Signal Processing*, pages 1–6, Los Alamitos. IEEE Computer Society.

[128] Lippmann, R. P. (1989). Pattern classification using neural networks. *IEEE Comunications Magazine*, 11:47–64.

[129] Lippmann, R. P. (1991). A critical overview of neural networks pattern classifiers. *Neural Networks for Signal Processing*, 30:266–275.

[130] Lourenço, P. M. (1998). *Um Modelo de Previsão de Curto Prazo de Carga Elétrica combinando Métodos Estatísticos e Inteligência Computacional.* D.Sc. thesis, Pontifícia Universidade Católica do Rio de Janeiro.

[131] Louzada-Neto, F. (1997). Extended hazard regression model for rehability and survival analysis. *Lifetime Data Analysis*, 3:367–381.

[132] Louzada-Neto, F. (1999). Modeling life data: A graphical approach. *Applied Stochastic Models in Business and Industry*, 15:123–129.

[133] Louzada-Neto, F. and Pereira, B. B. (2000). Modelos em análise de sobrevivência. *Ciência & Saúde Coletiva*, 8(1):8–26.

[134] Lowe, D. (1999). Statistics and neural networks advances in the interface. In *Radial Basis Function Networks and Statistics*, pages 65–95. Oxford University Press.

[135] Machado, M. A. S., Souza, R. C., Caldeva, A. M., and Braga, M. J. F. (2002). Previsão de energia elétrica usando redes neurais nebulosas. *Pesquisa Naval*, 15:99–108.

[136] Mackay, D. J. C. (1992). *Bayesian Methods for Adaptive Methods*. Ph.D. thesis, California Institute of Technology.

[137] Maksimovic, R. and Popovic, M. (1999). Classification of tetraplegics through automatic movement evaluation. *Medical Engineering and Physics*, 21(5):313–327.

[138] Mangiameli, P., Chen, S. K., and West, D. (1996). A comparison of SOM neural networks and hierarchical clustering methods. *European Journal of Operations Research*, 93(2):402–417.

[139] Mariani, L., Coradin, D., Bigangoli, E., Boracchi, P., Marubini, E., Pillot, S., Salvadori, R., Verones, V., Zucali, R., and Rilke, F. (1997). Prognostic factors for metachronous contralateral breast cancer: a comparison of the linear Cox regression model and its artificial neural networks extension. *Breast Cancer Research and Treatment*, 44:167–178.

[140] Markham, I. S. and Ragsdale, C. T. (1995). Combining neural networks and statistical predictors to solve the classification problem in discriminant analysis. *Decision Sciences*, 26:229–242.

[141] Martin, E. B. and Morris, A. J. (1999). Statistics and neural networks. In *Artificial neural networks and multivariate statistics*, pages 195–258. Oxford University Press.

[142] Mehrotra, K., Mohan, C. K., and Ranka, S. (1997). *Elements of Artificial Neural Networks*. MIT Press, Cambridge.

[143] Michie, D., Spiegelhalter, D. J., and Taylor, C. C., editors (2009). *Machine Learning, Neural and Statistical Classification*. Overseas Press, New Delhi.

[144] Mingoti, S. A. and Lima, J. O. (2006). Comparing SOM neural network with fuzzy c-means, K-means and traditional hierarchical clustering algorithms. *European Journal of Operational Research*, 174(3):1742–1759.

[145] Mitchell, T. M. (1997). *Machine Learning*. McGraw-Hill Inc., 2nd edition.

[146] Morettin, P. A. (1996). From Fourier to wavelet analysis of true series. In *Proceedings of Computational Statistics*, pages 111–122. Physica-Verlag.

[147] Morettin, P. A. (1997). Wavelets in statistics (text for a tutorial). In *The Third International Conference on Statistical Data Analysis Based on L_1-Norm and Related Methods*, Neuchâtel. (www.ime.usp.br/ pam/papers.html).

[148] Murtagh, F. (1994). Neural networks and related "massively parallel" methods for statistics: A short overview. *International Statistical Review*, 62(3):275–288.

[149] Neal, R. M. (1996). *Bayesian Learning for Neural Networks*. Springer, New York.

[150] Nelder, J. A. and McCullagh, P. (1989). *Generalized Linear Model*. CRC Press/Chapman & Hall, 2nd edition.

[151] Ng, S.-K. and McLachlan, G. J. (2007). Extension of mixture-of-experts networks for binary classification of hierarchical data. *Artificial Intelligence in Medicine*, 41(1):57–67.

[152] Nguyen, T., Malley, R., Inkelis, S. K., and Kupperman, N. (2002). Comparison of prediction models for adverse outcome in pediatric meningococcal disease using artificial neural network and logistic regression analysis. *Journal of Clinical Epidemiology*, 55:687–695.

[153] Nicole, S. (2000). Feedfoward neural networks for principal components extraction. *Computational Statistics and Data Analysis*, 33:425–437.

[154] Ohno-Machado, L. (1997). A comparison of Cox proportional hazard and artificial neural network models for medical prognoses. *Computers in Biology and Medicine*, 27:55–65.

[155] Ohno-Machado, L., Walker, M. G., and Musen, M. A. (1995). Hierarchical neural networks for survival analysis. In *The Eighth World Congress on Medical Information*, Vancouver.

[156] Orr, M. J. L. (1996). Introduction to radial basis function networks. Technical report, Institute for Adaptive and Neural Computation, Edinburgh University. (www.anc.ed.ac.uk/ mjo/papers/intro.ps).

[157] Ottenbacher, K. J., Smith, P. M., Illig, S. B., Limm, E. T., Fiedler, R. C., and Granger, C. V. (2001). Comparison of logistic regression and neural networks to predict rehospitalization in patients with stroke. *Journal of Clinical Epidemiology*, 54:1159–1165.

[158] Paige, R. L. and Butler, R. W. (2001). Bayesian inference in neural networks. *Biometrika*, 88(3):623–641.

[159] Park, Y. R., Murray, T. J., and Chen, C. (1994). Predicting sun spots using a layered perceptron neural network. *IEEE Transactions on Neural Networks*, 7(2):501–505.

[160] Patazopoulos, D., P, P. K., Pauliakis, A., Iokim-liossi, A., and Dimopoulos, M. A. (1998). Static cytometry and neural networks in the discrimination of lower urinary system lesions. *Urology*, 51(6):946–950.

[161] Patterson, D. W. (1998). *Artificial Neural Networks: Theory and Applications*. Prentice Hall, Upper Saddle River.

[162] Pedregosa, F., Varoquaux, G., Gramfort, A., Michel, V., Thirion, B., Grisel, O., Blondel, M., Prettenhofer, P., Weiss, R., Dubourg, V., Vanderplas, J., Passos, A., Cournapeau, D., Brucher, M., Perrot, M., and Duchesnay, E. (2011). Scikit-learn: Machine learning in Python. *Journal of Machine Learning Research*, 12:2825–2830.

[163] Peng, F., Jacobs, R. A., and Tanner, M. A. (1996). Bayesian inference in mixture-of-experts and hierarchical mixtures-of-experts models with an application to speech recognition. *Journal of the American Statistical Association*, 91:953–960.

[164] Pereira, B. B., Rao, C. R., Oliveira, R. L., and Nascimento, E. M. (2010). Combining unsupervised and supervised neural networks in cluster analysis of gamma-ray burst. *Journal of Data Science*, 8(2):327–338.

[165] Pereira, B. d. B. and R., R. C. (2005). Survival analysis neural networks. *Learning and Nonlinear Models - Revista da Sociedade Brasileira de Redes Neurais*, 2(2):50–60.

[166] Picton, P. (2000). *Neural Networks*. Palgrave Macmillan, London, 2nd edition.

[167] Pino, R., de la Fuente, D., Parreno, J., and Pviore, P. (2002). Aplicacion de redes neuronales artificiales a la prevision de series temporales no estacionarias a no invertibles. *Qüestiio*, 26:461–482.

[168] Poli, I. and Jones, R. D. (1994). A neural net model for prediction. *Journal of the American Statistical Association*, 89(425):117–121.

[169] Portugal, M. S. (1995). Neural networks versus true series methods: A forecasting exercise. *Revista Brasileira de Economia*, 49:611–629.

[170] Prybutok, V. R., Sanford, C. C., and Nam, K. T. (1994). Comparison of neural network to Shewhart $\bar{X}$ control chart applications. *Economic Quality Control*, 9:143–164.

[171] Qi, M. and Maddala, G. S. (1999). Economic factors and the stock market: A new perspective. *Journal of Forecasting*, 18:151–166.

[172] Rao, C. R. (1949). On some problems arising out of discrimination with multiple characters. *Sankya*, 9:343–365.

[173] Raposo, C. M. (1992). *Redes Neuronais na Previsão de Séries Temporais*. D.Sc. thesis, COPPE - Universidade Federal do Rio de Janeiro.

[174] Raudys, Š. (2001). *Statistical and Neural Classifiers: An Integrated Approach to Design*. Springer, London.

[175] Raza, M. Q. and Khosravi, A. (2015). A review on artificial intelligence based load demand forecasting techniques for smart grid and buildings. *Renewable and Sustainable Energy Reviews*, 50:1352–1372.

[176] Ripley, B. D. (1994). Neural networks and related methods for classification (with discussion). *Journal of the Royal Statistical Society B*, 56:409–456.

[177] Ripley, B. D. (1996). *Pattern Recognition and Neural Networks*. Cambridge University Press.

[178] Ripley, B. D. and Ripley, R. M. (1998). Artificial neural networks: Prospects for medicine. In *Neural Networks as Statistical Methods in Survival Analysis*. Landes Biosciences Publishing.

[179] Ripley, R. M. (1998). *Neural network models for breast cancer prognoses. D. Phill. Thesis*. Department of Engineering Science, University of Oxford. (www.stats.ox.ac.uk/ ruth/index.html).

[180] Rojas, R. (1996). *Neural Networks: A Systematic Introduction*. Springer.

[181] Rowe, D. B. (2003). *Multivariate Bayesian Statistics Models for Source Separation and Signal Unmixing*. Chapman & Hall/CRC.

[182] Rozenthal, M. (1997). *Estudo dos aspectos neuropsicológicos da esquizofrenia com uso de redes neurais artificiais*. D.Sc. in Psychiatry, Universidade Federal do Rio de Janeiro.

[183] Rozenthal, M., Engelhart, E., and Carvalho, L. A. V. (1998). Adaptive resonance theory in the search for neuropsychological patterns of schizophrenia. Technical Report in Systems and Computer Sciences, COPPE - Universidade Federal do Rio de Janeiro.

[184] Santos, A. M. (2003). *Redes Neurais e Árvores de Classificação Aplicadas ao Diagnóstico da Tuberculose Pulmonar Paucibacilar*. D.Sc. thesis, Universidade Federal do Rio de Janeiro.

[185] Santos, A. M., Seixas, J. M., Pereira, B. B., Medronho, R. A., Campos, M. R., and Calôba, L. P. (2005). Usando redes neurais artificiais e regressão logística na predição de hepatite a. *Revista Brasileira de Epidemiologia*, 8(2):117–126.

[186] Santos, A. M., Seixas, J. M., Pereira, B. B., Mello, F. C. Q., and Kritski, A. (2006). Neural networks: an application for predicting smear negative pulmonary tuberculosis. In *Advances in Statistical Methods for the Health Sciences: Applications to Cancer and AIDS Studies, Genome Sequence and Survival Analysis*, pages 279–292. Springer.

[187] Sarle, W. S. (1994). Neural networks and statistical models. In *Proceedings of the 19th Annual SAS Users Group International Conference*. (http//citeseer.nj.nec.com/sarle94neural.html).

[188] Sarle, W. S. (1996). Neural network and statistical jargon.

[189] Schumacher, M., Roßner, R., and Vach, W. (1996). Neural networks and logistic regression. Part I. *Computational Statistics and Data Analysis*, 21:661–682.

[190] Schwarzer, G., Vach, W., and Schumacher, M. (2000). On the misuses of artificial neural networks for prognostic and diagnostic classification in oncology. *Statistics in Medicine*, 19:541–561.

[191] Shamseldin, D. Y. and O'Connor, K. M. (2001). A non-linear neural network technique for updating of river flow forecasts. *Hydrology and Earth System Sciences*, 5:577–597.

[192] Silva, A. B. M., Portugal, M. S., and Cechim, A. L. (2001). Redes neurais artificiais e análise de sensibilidade: Uma aplicação à demanda de importações brasileiras. *Economia Aplicada*, 5:645–693.

[193] Silva, I. N. d., Hernane Spatti, D., Andrade Flauzino, R., Liboni, L. H. B., and dos Reis Alves, S. F. (2017). *Artificial Neural Networks: A Practical Course*. Springer International Publishing, Cham.

[194] Smith, A. E. (1994). X-bar and R control chart using neural computing. *International Journal of Production Research*, 32:277–286.

[195] Smith, M. (1993). *Neural Network for Statistical Modelling*. Van Nostrand Reinhold.

[196] Specht, D. F. (1991). A general regression neural networks. *IEEE Transactions on Neural Networks*, 2(6):568–576.

[197] Specht, D. F. (1998). Probabilistic neural networks for classification, mapping, or associative memories. In *Proceedings of International Conference on Neural Networks*, volume 1, pages 525–532.

[198] Stegemann, J. A. and Buenfeld, N. R. (1999). A glossary of basic neural network terminology for regression problems. *Neural Computing & Applications*, 8(4):290–296.

[199] Stern, H. S. (1996). Neural network in applied statistics (with discussion). *Technometrics*, 38:205–220.

[200] Stone, J. V. (2004). *Independent Component Analysis: A Tutorial Introduction*. MIT Press, Cambridge.

[201] Stone, J. V. and Porril, J. (1998). Independent component analysis and projection pursuit: a tutorial introduction. Techinical Report, Psychology Departament, Sheffield University.

[202] Sun, K. T. and Lai, Y. S. (2002). Applying neural netorks technologies to factor analysis. In *Procedings of the National Science Council, ROD(D)*, volume 12, pages 19–30.

[203] Swanson, N. R. and White, H. (1995). A model-selection approach to assessing the information in the term structure using linear models and artificial neural networks. *Journal of Business & Economic Statistics*, 13(3):265–75.

[204] Swanson, N. R. and White, H. (1997). Forecasting economic time series using flexible versus fixed specification and linear versus nonlinear econometric models. *International Journal of Forecasting*, 13(4):439–461.

[205] Terasvirta, T., Dijk, D. V., and Medeiros, M. C. (2005). Linear models, smooth transitions autoregressions and neural networks for forecasting macroeconomic time series: a re-examination. *Internation Journal of Forecasting*, 21:755–774.

[206] Terasvirta, T., Lin, C. F., and Granger, C. W. J. (1993). Power properties of the neural network linearity test. *Journal of Time Series Analysis*, 14:209–220.

[207] Tibshirani, R. (1996). A comparison of some error estimates for neural networks models. *Neural Computation*, 8:152–163.

[208] Titterington, D. M. (2004). Bayesian methods for neural networks and related models. *Statistical Science*, 19(1):128–139.

[209] Tkacz, G. (2002). Neural network forecasting of Canadian GDP growth. *International Journal of Forecasting*, 17:57–69.

[210] Tseng, F.-M., Yu, H.-C., and Tzeng, G.-H. (2002). Combining neural network model with seasonal time series arima model. *Technological Forecasting and Social Change*, 69(1):71–87.

[211] Tu, J. V. (1996). Advantages and disadvantages of using artifical neural networks versus logit regression for predicting medical outcomes. *Journal of Clinical Epidemiology*, 49:1225–1231.

[212] Tura, B. R. (2001). Aplicações de data mining em medicina. M.Sc. thesis, Faculdade de Medicina, Universidade Federal do Rio de Janeiro.

[213] Vach, W., Robner, R., and Schumacher, M. (1996). Neural networks and logistic regression. Part II. *Computational Statistics and Data Analysis*, 21:683–701.

[214] Venkatachalam, A. R. and Sohl, J. E. (1999). An intelligent model selection and forecasting system. *Journal of Forecasting*, 18:167–180.

[215] Vuckovic, A., Radivojevic, V., Chen, A. C. N., and Popovic, D. (2002). Automatic recognition of alertness and drowsiness from EEC by artificial neural networks. *Medical Engineering and Physics*, 24(5):349–360.

[216] Warner, B. and Misra, M. (1996). Understanding neural networks as statistical tools. *American Statistician*, 50:284–293.

[217] Weller, S. and Rommey, A. K. (1990). *Metric Scalling: Correspondence Analysis*. Number 75 in Quantitative Applications in Social Science Series. SAGE Publications.

[218] West, D. (2000). Neural networks credit scoring models. *Computers and Operations Research*, 27:1131–1152.

[219] White, H. (1989a). Learning in artificial neural networks: A statistical perspective. *Neural Computation*, 1(4):425–464.

[220] White, H. (1989b). Neural-network learning and statistics. *AI Expert*, 4(12):48–52.

[221] White, H. (1989c). Some asymptotic results for learning in single hidden-layer feedforward network models. *Journal of the American Statistical Association*, 84(408):1003–1013.

[222] White, H. (1996). Parametric statistical estimation using artificial neural networks. In *Mathematical Perspectives on Neural Networks*, pages 719–775. L Erlbaum Associates.

[223] Wilde, D. (1997). *Neural Network Models*. Springer, 2nd edition.

[224] Wold, H. (1938). *A Study in the Analysis of Stationary Time Series*. Almquist & Wiksells, Stockholm.

[225] Wu, C. H. and McLarty, J. W. (2000). *Neural Networks and Genome Informatics*. Elsevier.

[226] Xiang, A., Lapuerta, P., Ryutov, Buckley, J., and Azen, S. (2000). Comparison of the performance of neural networks methods and Cox regression for censored survival data. *Computational Statistics and Data Analysis*, 34:243–257.

[227] Xiao, H., Rasul, K., and Vollgraf, R. (2017). Fashion-MNIST: a novel image dataset for benchmarking machine learning algorithms.

[228] Yang, Z., Platt, M. B., and Platt, H. D. (1999). Probabilistic neural networks in bankruptcy prediction. *Journal of Business Research*, 44(2):67–74.

[229] Zhang, G., Patuwo, B. E., and Hu, M. Y. (1998). Forecasting with artificial neural networks: the state of the art. *International Journal of Forecasting*, 14:35–62.

[230] Zhang, G. P. (2007). Avoiding pitfalls in neural network research. *IEEE Transactions on Systems, Man, and Cybernetics - Part C: Applications and Reviews*, 37(1):3–16.

[231] Zhang, Q. and Benveniste, A. (1992). Wavelets networks. *IEEE Transactions on Neural Networks*, 3(6):889–898.

Index

Printed in the United States
by Baker & Taylor Publisher Services